The Journey
of Man

The Journey of Man
A Genetic Odyssey

Spencer Wells

Photographs by Mark Read

Princeton University Press

Princeton and Oxford

Published in the United States and Canada by Princeton University Press,
41 William Street, Princeton, New Jersey 08540

In the United Kingdom and European Union, published by the Penguin Group
Penguin Books Ltd, 80 Strand, London WC2R 0RL

Library of Congress Control Number 2002113609

ISBN 0-691-11532-X

This book has been composed in 10.5/14 pt PostScript Adobe Sabon

Printed on acid-free paper. ∞

www.pupress.princeton.edu

Printed in the United States of America

10   9   8   7   6   5   4

To my wife, Trendell, and our daughters, Margot and Sasha
(Y-chromosomes are overrated anyway . . .)

The aim of science is not to open the door to infinite wisdom, but to set a limit to infinite error.
Bertolt Brecht, *Life of Galileo*

Contents

Maps

Figures

Most of us can name our grandparents, many our great-grandparents, and some our great-great-grandparents. Beyond that, we enter a dark and mysterious realm known as history, through which we can only navigate with hesitant steps, feeling our way with whispered guidance. Who were the people that came before? Where did they live? What were their lives like?

In this book I will argue that the answers to these questions can be found in our genetic code, which makes us uniquely human – but also makes us unique individuals. Our DNA carries, hidden in its string of four simple letters, a historical document stretching back to the origin of life and the first self-replicating molecules, through our amoebic ancestors, and down to the present day. We are the end result of over a billion years of evolutionary tinkering, and our genes carry the seams and spot-welds that reveal the story.

It is not the code itself that delivers the message, but rather the differences we see when we compare DNA from two or more individuals. These differences are the historical language of the genes. In the same way that you wouldn't include 'water-dwelling' in a classification of fish, because all fish live in water, the identical bits of our genetic code tell us nothing about our history. The story is in the differences, and this is what we study.

This is not a book on human origins. Rather, it is about the journey we have taken as a species, from our birthplace in Africa to the far corners of the earth, and from the earliest evidence of fully modern humans to the present day – and beyond. The argument pursued throughout is that genetics provides us with a map of our wanderings and gives us a rough idea of the dates – and it is up to us to reconcile this data with the archaeological and climatological record in order to fill in the picture. Of course, every journey must have a beginning, and this one is no exception. It begins with the scientific effort to make sense of human diversity, which leads us to the birthplace of our species. The methods we use to infer our African origins are the same ones we then use to trace humanity's global journey. It is the journey itself that is the main focus, and for this reason most of the details of our hominid ancestors have been left out.

This book was originally conceived as part of a documentary film project of the same name. It has since taken on a life of its own, and it stands alone as a unique entity, providing greater detail on the scientific background than is possible on television. The film, on the other hand, presents an (almost) first-hand experience, conveying the excitement and adventure of the journey in a way that only moving images can, so I hope that readers of this book will enjoy it equally.

Although it was often difficult to juggle the film and the book at the same time, this did provide some significant advantages. For me, the chance to retrace my own personal 'journey of man' and to meet people from around the world – to see how they live and to discuss the scientific results with them – were profound and wonderful experiences. I hope that a sense of this has come through in print.

The title was chosen for a reason – and it wasn't sexism. The journey we trace is primarily one made by men, because it is the Y-chromosome, inherited from Adam down the male line, which gives us our keenest tool for deciphering the journey. The Y helps to place the stones, bones and languages in context better than any other part of our genetic code, and ultimately gives us the genetic answers we are looking for. Of course, in order to leave descendants, these early human groups must have included women; while the journey we follow may leave out some female-specific details, the resolution we can achieve only by following the male lineage is worth the omission.

What follows is a scientific detective story guided by the temporal order of events. We begin with a deceptively simple question: how do we decide if the concept of human 'race' has any validity? Are we indeed all one species, or are there discrete divisions among human groups? After all, we appear to be so different from one another. The answer to this question, first provided by my PhD adviser at Harvard,

Richard Lewontin, gives us a clue about the journey – but it doesn't reveal the crucial details.

The second main question concerns our geographic distribution. How did we come to occupy every corner of the globe? The DNA markers are able to provide us with the details. The methods for doing this, developed over the course of half a century, have been greatly influenced by Luca Cavalli-Sforza, with whom I was lucky enough to work as a postdoctoral fellow at Stanford in the 1990s. It was Luca's insight, as a geneticist with a passion for history and a talent for mathematics, which provided us with a time machine capable of resurrecting the stories of the past from people living in the present. This book could not have been written without his intellectual presence, and it is impossible not to feel humbled while taking in the view from his shoulders.

One of the most compelling things about being on an archaeological dig is the sense that you are actually seeing and handling implements that were last touched by human hands hundreds or thousands of years ago. Often the sense is so great that a feeling of *déjà vu* comes over you, and it seems as though you have somehow been transported back in time. When, as a child, I saw the Tutankhamun exhibition that toured the United States, I was struck by the combination of modern skill and ancient subject matter. It seemed as though the pieces, although fabulously exotic, could have been made the week before by a skilled craftsman. The fact that they were nearly 4,000 years old was extraordinary, and sparked in me a curiosity for the past that has never diminished.

Genetics, at least the branch of it that informs the subject of human origins and migrations, is necessarily less visually compelling than archaeology, despite the fascinating stories it tells. Mark Read's wonderful photographic portraits included in the book go a long way towards correcting this imbalance. I have been lucky enough to work with Mark on several occasions, both in the course of sample-collecting expeditions in Asia and during the making of the film. He is a talented and dedicated artist, and his work adds enormously to the book. Mark's photographs reflect the way people actually live today, because our knowledge of genetic history is inferred from the blood of people living in the present – it is their *living* genomes that give us our clues.

Every one of us is carrying his or her personal history book around inside us – we simply need to learn how to read it.

The Australian Aborigines maintain their sense of connection to their ancestors and their homeland through a musical story – something Bruce Chatwin and others have called a 'songline'. These songlines reflect the actual journey taken by their ancestors during the Dreamtime, a period in the distant past, before collective memory. In a sense, this is precisely what we are trying to do with our studies of DNA – resurrect a global songline for everyone alive today, describing how they reached their current location and what the journey was like. As secular Westerners we have lost our traditional songlines to a greater extent than other peoples around the world, so it is perhaps appropriate that Western science has developed the methods for rediscovering them. However, our research does not take place in a vacuum, and science can sometimes run roughshod over cultural beliefs. I would hope that this book might be a small step towards changing the field into what it really is – a collaborative effort between people around the world who are interested in their shared history.

So, with that as an overview, let's start our genetic excavation. The past awaits . . .

1
The Diverse Ape

So God created man in his own image, in the image of God created he him; male and female created he them. And God blessed them, and God said unto them, Be fruitful, and multiply . . .
Genesis 1: 27–8

Creation myths can be found at the core of all religions. Most seek to answer the child's question 'where do we come from?' – to explain our existence and our place in the world in a succinct way. But while they may attempt to explain how we originated, creation myths fail to account for the spectrum of cultures, shapes, sizes and colours we see when we look at people around the world. Why do we look so different from each other, and how did we come to inhabit such far-flung places?

Herodotus, the fifth-century BC Greek historian, provided posterity with far more than a description of the Greco-Persian wars. He also gave us our first clear descriptions of human diversity, viewed through an idiosyncratic classical lens. We learn of the dark and mysterious Libyans, the barbaric man-eating Androphagi of the Russian north, and hear descriptions of people who seem to resemble the Turks and Mongolians. Herodotus relates fanciful tales of griffins guarding precious hoards in the mountains of Asia, and we are treated to exotic descriptions of tribes in northern India who collect gold from the burrows of ants. Overall, it is a tour de force – the first ethnographic treatise in Western literature and, despite its obvious flaws, a valuable snapshot of the known world at that time.

If we were to assume the role of a naïve modern-day Herodotus and fly an equatorial route around the world, the diversity of people and places would be astounding. Imagine for a moment being on a plane above the Atlantic Ocean at the very centre of the Cartesian globe, 0° longitude, 0° latitude – about 1,000 km west of Libreville, Gabon, in west-central Africa. If we imagine the plane flying east, and allow ourselves the science-fiction trick of being able to scan the ground from our vantage point in the sky, we will get a small sample of humanity's diversity.

The first people we encounter are Africans – specifically, central Africans, speaking Bantu languages. They have very dark skin, and live primarily in small villages hacked from the forest. As we move further east, we still see dark-skinned people, but these look somewhat different. They are the tall, thin Nilotic peoples of east Africa – some of the tallest on earth. They live on grassy savannahs, and are almost completely dependent on their cattle for survival. Scattered in amongst these groups are people who speak yet another language – one which is as different from Nilotic and Bantu as they are from each other, even though they live close by – the Hadza.

As we continue east we encounter a huge body of water – so vast that it is impossible to see across it, and it seems an eternity before we reach an archipelago known as the Maldives. The people here seem quite different from those we saw in Africa, and speak yet another language. Their skin is dark, like that of the Africans, but their faces are different – nose shape, hair type and other minor details. They are clearly related to the Africans, but differ in obvious ways.

As we continue on our journey – above the same enormous body of water – we see a large island rising up ahead of us. We have reached Sumatra, and here we encounter yet another type of human, somewhat smaller than the Africans and peoples of the Maldives, with yet another facial appearance – very straight hair, lighter skin and a thicker layer of skin above the eyes. Further east, passing countless other islands, we again encounter people with very dark skin, known as Melanesians. They are unlike the Africans in many other ways, so is their dark skin a characteristic that evolved in this region? Or is it indicative of a close connection with Africa?

Next we encounter the Polynesians, living on small coral atolls separated by thousands of miles of open ocean. They appear to be somewhat similar to the Sumatrans encountered before but, as always seems to be the case, they are different. The biggest question is why they are living in such remote locations – how did they get there?

Continuing on our route, we encounter the coast of Ecuador, in western South America. In the capital, Quito, we find an odd mix of people. There seem to be two main types: those who in some respects resemble the peoples of the Maldives, but with lighter skin, and those who in many ways resemble the Sumatrans and Polynesians. It seems odd to find such divergent types of humanity living in the same place, since the other locations we have visited tended to be more homogeneous. Why is Ecuador different? A disparate mix of people is found further east on the continent, where on the north-eastern coast of Brazil we encounter Africans again – but living far from Africa! During the long journey back to our starting point we ponder the tapestry we have just seen, and try to formulate an explanation for the pattern of diversity.

Our short tour of the world was a kind of thought experiment, where we imagine what it must have been like to encounter things as people may have done a few hundred years ago, during the first European 'voyages of discovery'. By assuming the guise of ignorance, we can ask simple questions that seem trivial to us today, given our knowledge of history. The interesting thing about this thought experiment is that, until very recently – excepting the Africans and Europeans encountered in South America – there was no ready explanation for the patterns we saw.

One species . . .

On 30 June 1860 an angry cleric named Samuel Wilberforce mounted the stage at Oxford University's Museum Library. He was primed for a fight – not just for himself, but for something far more important: his worldview. Wilberforce felt that he was fighting for the future of Christianity. The venue was a formal debate on the place of man in nature, a field of enquiry until recently limited to philosophers and the church. The good bishop, taking scripture at its most literal, believed the world to be around 6,000 years old, created by the hand of God on 23 October 4004 BC, a date obtained by counting back through the genealogy described in the Bible. In his speech he asked a pointed question – one that was on the minds of many in the audience. Was it really possible that he could be related to a monkey? It sounded so preposterous!

Wilberforce was a polished speaker, and to many in the audience his argument was persuasive. But while he held his own in the library that day, in the long run he was destined to be trounced. And, foreshadowing a significant change in the way we viewed our place in the world, the dragon slayers were not philosophers or clergymen but professional scientists. Joseph Hooker and Thomas Henry Huxley, both Victorians par excellence, were strong supporters of Charles Darwin's new theory of evolution by natural selection. Huxley, lecturer in biology at the London School of Mines, later became better known as 'Darwin's Bulldog'. Hooker was an accomplished botanist and assistant director of the Royal Botanical Gardens at Kew. When they rose at the end of Wilberforce's lecture to refute his emotional arguments, they were sounding a death-knell to the old views on human origins. Science was leading the way into a brave new world.

The debate between Wilberforce, Hooker and Huxley served not merely to reinforce the public's acceptance of evolution – most educated people had already come to see the world in an evolutionary context – but rather to realign humanity's place in it. When we viewed ourselves as the divine creation of an omnipotent being, we could easily justify our isolation from the rest of the living world. Masters, conquerors, perhaps favoured children – but different.

Darwin's insight had changed all of that. This dyspeptic near-recluse had, with a few strokes of his pen (and some twenty years of dabbling with pigeons and barnacles), demoted humanity from divine creation to a product of biological tinkering. And the odd thing is that he hadn't even set out to do this. Darwin, the scion of a wealthy Victorian family (his grandfather was Josiah Wedgwood, his father was a wealthy physician, and Darwin himself spent part of each day looking after his investments), had no intention of rocking the boat when he set out on his voyage of discovery aboard the *Beagle* in 1831. He was certainly looking for adventure, and needed something to stave off the looming inevitability of a staid country parsonage – the logical career choice for a Cambridge graduate of that era. But he was looking for something else as well.

As was the case with many Victorians, Darwin had developed a keen interest in science during his childhood. While he had the usual chemistry accidents, especially with his older brother Erasmus – with whom he once destroyed a garden shed-cum-laboratory when an experiment went explosively awry – Darwin's interests were primarily of the outdoor variety. He was inordinately fond of beetles (he once wrote in a letter of 'pining' for a like-minded beetle fancier), and spent many hours in the field scavenging for exotic specimens. But it was his interest in geology, developed while he was a student at Cambridge, that was to have the greatest impact on his future work.

Geology was undergoing a revolution in the early nineteenth century – one which was calling into question our whole understanding of history, as handed down in the Bible. Darwin was an adherent of a school of thought that became known as uniformitarianism, first formulated by Charles Lyell. Lyell believed that the forces and materials found in the world today had always behaved in essentially the same way – even in the distant past. Diametrically opposed to the uniformitarian school were the catastrophists – led by major scholars such as Louis Agassiz, a Swiss transplant to America who founded Harvard University's Museum of Natural History. The catastrophists believed that the earth went through long periods of stasis when nothing much happened, but that occasionally all hell would break loose. This could take the form of a biblical flood, or an ice age, or a massive upheaval in the earth's crust. All major changes – in organisms as well as the planet itself – were driven by these freak events. The distribution of the world's plant and animal species was due to a series of catastrophic events during their history.

The problem with catastrophism was that it relied too much on odd happenings to be of any use – there were rather a lot of changes that seemed to have occurred without any drastic catalysts. If change could occur without invoking a major causal event, then why was it necessary to invoke them at all? Why not simply assume that the earth is constantly changing at a very gradual rate, and that over long periods of time these incremental steps produce significant results? It seemed so much easier to reconcile with the actual data, said Lyell.

All of this was percolating in young Darwin's mind when he set out aboard HMS *Beagle*, engaged as a 'gentleman companion' for Captain FitzRoy. This unusual position had to do with Victorian social customs, in that the Captain was considered to be of too high a social class to mix with the crew. There was, in fact, an official naturalist on board the ship – the ship's surgeon – but he ended up leaving the voyage in Brazil after a falling out with FitzRoy. At any rate, Darwin was the *de facto* naturalist on the journey, and his lack of official status as such allowed him enormous leeway in pursuing his own studies.

His journal from the five-year journey, *The Voyage of the Beagle*, is a classic of nineteenth-century travel literature. During the trip, Darwin made several major discoveries, including finding a reasonable explanation for why coral atolls are round (it has to do with receding volcanoes) and deciding that the Tahitians were very attractive people indeed. The most important – his initial insight into the action of natural selection, and its role in the origin and evolution of species – has been examined so often that it isn't necessary to reiterate here. Suffice to say, Huxley and Wilberforce would never have faced off in 1860, and you wouldn't be reading this book, if Darwin hadn't recognized natural selection as the driving force of evolution.

It is one of Darwin's other subjects, discernible even in this, his earliest major work, which interests us here. It is a subject which is dealt with more subtly than his discussion of biological evolution, presaging his hesitation nearly thirty years later to include a direct discussion of it in *The Origin of Species*. The subject is humanity. Or rather, the diverse array of humanity encountered through the lens of a Victorian scientist with an urge to explain the patterns he saw. Why were people around the world so different from each other?

The *Beagle* set sail from Devonport, near Plymouth, on 27 December 1831, calling at the Cape Verde Islands, Brazil, Argentina, Tierra del Fuego, Chile, Ecuador, the Galapagos, Tahiti, New Zealand, Australia, Mauritius and Brazil (again) before returning home on 2 October 1836. Travelling on such a grand, circuitous route, Darwin had a chance to encounter many different groups of people first-hand. He explored Brazil, witnessed the gauchos of Argentina in action on the pampas and trekked into the Andes with Chilean guides. Perhaps the most distinctive people he encountered, though, were the native inhabitants of Tierra del Fuego.

Darwin described the Fuegians as being '. . . stunted in their growth, their hideous faces bedaubed with white paint, their skins filthy and greasy, their hair entangled, their voices discordant, and their gestures violent. Viewing such men, one can hardly make one's self believe that they are fellow-creatures . . .' Clearly not what most people conjure up when asked to describe 'noble savages'. Yet Darwin was actually travelling with three Fuegians taken to London five years earlier by Captain FitzRoy. Colourfully named Fuegia Basket, Jemmy Button and York Minster by their kidnappers, their real names were Yok-cushlu, Orundellico and El'leparu. Taken by the sailors on the earlier voyage as a form of ransom after a small boat was stolen, the Fuegians were clearly out of their element in the world of Victorian Britain. Nevertheless, they had learned to speak rudimentary English and had even begun to take on some of the affectations of the British middle classes. Jemmy, for instance, repeatedly exclaimed 'poor, poor fellow' when Darwin was seasick – which he was with disheartening regularity. In spite of the alien nature of the Fuegians in their native land, Darwin clearly views them as being members of the same species, albeit with his Victorian class-influenced view of humanity. He even compares them favourably with the sailors on the *Beagle* when discussing super-stitions, and blames their generally lower level of material culture on an egalitarian political system. Although he may have been rather naïve politically, he was ahead of his time scientifically.

Importantly, Darwin had come down on the side of nurture in the nature vs. nurture debate. Even the Fuegians, as horrendous as they were in their natural state, were members of the same species as the crew aboard the *Beagle*. In the closing chapter of his journal he takes a jab at the barbaric slave trade then widespread in the Americas with one of the most poignant statements ever made on the equality of humanity: 'It is often attempted to palliate slavery by comparing the state of slaves with our poorer countrymen: if the misery of our poor be caused, not by the laws of nature, but by our institutions, great is our sin . . .'

But if humans were all members of the same species, how was it possible to explain the dizzying diversity in human colours, shapes, sizes and cultures around the world? Where had the species originated – and how had our ancestors journeyed to such remote parts as Capetown, Siberia and Tierra del Fuego? The answers to these questions would need to wait another 150 years, with a few detours through bones, blood and DNA.

. . . or many?

How do we define a species? The accepted definition since the mid-twentieth century is that of an interbreeding (or potentially inter-breeding, in the case of widely dispersed species) group of organisms. In other words, if it is possible to reproduce young together, you must be the same species. To Darwin, writing before the acceptance of this codified definition, there nonetheless seemed to be no question as to the commonality of humanity. His abolitionist call at the end of the *Voyage* was heart-felt, as slavery had recently been outlawed in Britain, and the debate still raged in the United States and elsewhere. But many others had taken quite a different view, arguing vehemently that humanity was clearly divided into distinct species or subspecies. This was first formalized in the early eighteenth century by a Swedish botanist, Carl von Linne (Latinized to Linnaeus), who took it upon himself to classify every living organism on the planet. Rather a daunt-ing task, but Linnaeus managed to do a pretty good job. Among other innovations, he gave us the binomial system of nomenclature used by biologists to this day – the Latin *Genus species* we all know from school, as in *Homo sapiens*.

Linnaeus recognized that all humans were part of the same species, but he added additional subclassifications for what he saw as the races, or subspecies, of humanity. These included *afer* (African), *ameri-canus* (Native American), *asiaticus* (east Asian), and *europaeus* (Euro-pean), as well as a poorly defined, blatantly racist category he called *monstrosus* – which included Darwin's Fuegians, among other groups. To Linnaeus, it seemed that the differences among humans were great enough to warrant this additional classification.

Darwin, ever the objective scientist, noted that our outward appear-ance has been over-emphasized in classifying humanity. In *The Descent of Man*, written towards the end of his life, he notes that: 'In regard to the amount of difference between the races, we must make some allowance for our nice powers of discrimination gained by long habit of observing ourselves.' This is an important insight, and one that helps to explain much of the subsequent debate over human origins.

The American pro-slavery lobby embraced an extreme form of the Linnean view in the nineteenth century. The view that human races were actually separate, inherently unequal entities made it easier to justify the brutal oppression practised in the United States. The theory that human races are distinct entities, created separately, is known as polygeny – from the Greek for 'many origins'. This theory clearly contradicted the biblical story of the Garden of Eden, inhabited by a single Adam and a single Eve, and thus raised the hackles of the church. Most biologists also objected to the polygenist view, noting the extensive hybridization among human races. To the polygenists these objections were easily overcome, as exemplified by Louis Agassiz, our Swiss catastrophist. According to Stephen Jay Gould, Agassiz believed that the ancients who wrote the Bible would not have been familiar with the different types of humanity, and thus they only wrote about a Mediterranean Adam. Agassiz thought that the Negroid Adam must have existed, as well as the Mongolid, and presumably the American.

While most biologists did not accept this view, it has been maintained to the present day in some anthropological literature. This is largely as a result of the great difficulty in explaining the physical diversity in humans, as well as certain patterns in the fossil record. Perhaps the best-known recent adherent of this view was the American anthropologist Carleton Coon, who published two hugely influential books in the 1960s, *The Origin of Races* and *The Living Races of Man*. In these books, Coon advanced the theory that there are five distinct human subspecies (Australoid, Capoid, Caucasoid, Congoid and Mongoloid), which evolved into their present forms *in situ* from ancestral hominids. Tellingly, Coon suggests that the different subspecies evolved at different times, with the African Congoids appearing early and remaining trapped in an evolutionary dead-end until the present. He asserts that the dominance of the Europeans is a natural consequence of their evolved genetic superiority, and even provides solace for those who lie awake at night worrying about interracial mixing:

Racial intermixture can upset the genetic as well as the social equilibrium of a group, and so, newly introduced genes tend to disappear or be reduced to a minimum percentage unless they provide a selective advantage over their local counterparts. I am making these statements not for any political or social purpose but merely to show that, were it not for the mechanisms cited above, men would not be black, white, yellow or brown.

This was not a statement to be taken lightly, considering that the writer was the president of the American Association of Physical Anthropology (the largest and most influential anthropological organization in the world), a professor at the University of Pennsylvania, curator of ethnology at the University Museum and a regular guest on a popular American television programme.

It is interesting that Coon went to such an effort to distance himself from political motivations. He did this because physical anthropology was just emerging from a dark period when it had, in fact, been self-consciously political. As outlined by one of its main proponents, Aleš Hrdlička, in the inaugural issue of the *American Journal of Physical Anthropology* in 1917, physical anthropology should serve humanity as well as study it – it was not simply a 'pure' science. Hrdlička noted its utility in formulating eugenics programmes, as well as in determining immigration policy. While he may have been trying to impress funding agencies with the applicability of what many considered to be a rather esoteric science, it was clear that some people were listening quite closely – and were soon to act on the advice of some pragmatic and politically savvy anthropologists.

Out of the ivory tower

Anthropology had developed in the nineteenth century as 'that science which has for its object the study of mankind as a whole, in its parts, and in relation with the rest of nature'.

More than anyone else, the Frenchman Paul Broca – who penned this description – can be credited with creating the modern discipline of physical anthropology. Broca was an expert on craniometry, the measurement of minute differences in skull morphology that were thought by some to indicate innate potential, and he developed a detailed classification of humanity based on these subtle variations. Broca's methods, disseminated in a highly influential textbook, served to galvanize the scientific community. Soon everyone wanted to measure skulls.

In England, an amateur scientist named Francis Galton was an early convert of Broca's. Galton had inherited enough money to fund a variety of research subjects, including statistics and biology. Soon he too began to measure anything and everything on the human body in an effort to categorize human diversity scientifically. This could have been dismissed as no more than an eccentric's dabblings had his fascination with human classification not mixed with a misinterpretation of Darwin's theory of natural selection to produce a potent brew.

As we saw earlier, Darwin was not a 'hard' racist. He was as prone to trivial biases as the next person, but from his few statements on the subject, we can infer that he believed humanity to be largely equivalent in terms of its biological potential. This was not true for many of his adherents. It was the philosopher Herbert Spencer, for instance, who actually coined the phrase 'survival of the fittest', and he used it to justify the social divisions inherent in late-nineteenth-century Britain in a series of widely read books and essays. If divisions within society could be explained by science, then surely differences between cultures had a similar cause. Combined with the Victorian obsession with classification, this leap from 'might makes right' to a belief that these cultural differences must be definable using scientific methods encouraged the growth of the eugenics movement.

The movement began innocently enough. Eugenics actually means 'good birth' (who could oppose that?) and, to a certain extent, it had always existed. The collection of ancient Jewish laws known as the Talmud, for instance, urged men to sell all their possessions in order to afford to marry the daughter of a scholar, so that their children would be more intelligent (scholars' daughters clearly weren't cheap dates). It was only at the end of the nineteenth century, however, that eugenics really took off. The reasons are complex and have to do with Victorian ideas of self-improvement, interest in new scientific fields such as genetics and the wealth of emerging data from physical anthropology. Once it got going, though, there was no stopping it.

The Eugenics Education Society was founded Britain in 1907, in Galton's honour. Its stated objective was to improve the gene pool of humanity through the selective breeding of 'fit' individuals. Its influence spread rapidly to the United States, where the culture was particularly predisposed to theories that promised self-improvement through the application of scientific knowledge. Soon 'Fitter Families' contests were a common feature of American state fairs, with families vying for the kudos and medals that came with being chosen as the fittest. Eugenics also caught on throughout Europe, where a somewhat darker strain emerged in the form of German racial hygiene.

While eugenics began as a movement dedicated to social enlightenment, its aims were soon perverted, and by the 1910s and 20s it was being used in the United States as scientific justification for the forced sterilization of people believed to be mentally subnormal. It was also behind the mean-spirited implementation of racist immigration tests and quotas (in the 1920s desperately poor eastern European immigrants, most of whom were illiterate, were expected to arrive at Ellis Island in New York knowing how to read). The systematic extermination of Jews, gypsies, homosexuals and other supposedly inferior groups by the Nazis in the 1940s had its scientific justification in the application of eugenic principles. Physical anthropology had jumped to the head of the queue in its race to prove 'useful'.

It is no wonder then that Coon, writing after the horrible truth about the Nazi atrocities had come to light, made such an effort to distance himself from political ends. Even in the segregationist climate of America in the early 1960s he would have inflamed old wounds that were only beginning to heal if he had recommended political action based on the findings of physical anthropology. Instead, he presented the fact of human racial differences as an objective, scientific observation of the world – warts and all. Don't blame the messenger, he seemed to be saying, if you don't like the message. But the claim that his conclusions were based on an objective appraisal of the evidence at hand was flawed, since no one had actually tested his genetic hypotheses. What *did* our genes have to say about human racial differences?

2
E pluribus unum

What we call the beginning is often the end
And to make an end is to make a beginning.
The end is where we start from.
T. S. Eliot, 'Little Gidding' (*Four Quartets*)

The study of human diversity was, until the twentieth century, limited to variation that could be observed with the naked eye. The subject of countless studies by Broca, Galton and the biometricians in Europe and America, this era marked a 'collection' phase of physical anthropology – the early stages of a new field of scientific enquiry, when there is no unifying theory with which to analyse the data accumulated. There was only one problem with the growing mass of data on human morphological variation – there was no simple correspondence between the newly rediscovered laws of heredity and the characters being measured. While there is certainly a genetic component to human morphology, it is clear that dozens – probably hundreds – of separate genes control this variability. Even today, the underlying genetic causes have yet to be deciphered. Thinking of Broca's craniometric studies, if a particular bump on the skull is found in two unrelated individuals, does it necessarily represent the same genetic change? Are the bumps really the same characteristic, and thus representative of a true genetic relationship, or do they simply resemble each other superficially – by chance? It was impossible to know.

Genetic variation was critical for the study of human diversity because is it is genetic change that actually produces evolution. At its most basic level, evolution is simply a change in the genetic composition of a species over time. Thus in order to assess how closely related individuals are – in particular whether they form a single species – it is important to know something about their genes. If the genes are the same, then they are the same species. What physical anthropology desperately needed was a collection of varying traits – known as polymorphisms, from the Greek for 'many forms' – with a simple pattern of inheritance. These could then be used to study human diversity in an effort to categorize it. Some traits like this were already known, particularly diseases like haemophilia. The problem with disease-causing polymorphisms was that they were simply too rare to be of any use in classification. Common, genetically simple polymorphisms were critical.

These arrived in 1901, when Karl Landsteiner noticed an interesting reaction upon mixing the blood from two unrelated people: some of the time it clumped together, forming large clots. This coagulation reaction was shown to be heritable, and it constituted the first demonstration of biochemical diversity among living humans. This experiment led to the definition of human blood groups, which would soon be applied to transfusions all over the world. If your doctor tells you that you have type A blood, this is actually the name given by Landsteiner to the first blood group polymorphism over a century ago.

Building on Landsteiner's insight, a Swiss couple named Hirszfeld began to test the blood of soldiers in Salonika during the First World War. In a 1919 publication, they noted different frequencies of blood groups among the diverse nationalities thrown together by the hostilities – the first direct survey of human genetic diversity. The Hirszfelds even formulated a theory (accepted by some to this day) in which the A and B blood groups represent the traces of 'pure' populations of aboriginal humans, each composed entirely of either A or B individuals. These pure races later became mixed through migration, leading to the complicated patterns of A and B seen in their study. They failed to explain how the two races may have arisen, but given that group A was thought to have originated in northern Europe, while B was a southern marker at highest frequency in India, it seems that there must have been two independent origins of modern humans.

In the 1930s an American named Bryant and an Englishman named Mourant, building on the work of the Hirszfelds, began to test blood samples from around the world. Over the next thirty years these two men and their colleagues would examine thousands of people, from hundreds of populations, both living and dead. Bryant and his wife (like the Hirszfelds, another of the marital duos in population genetics) even went so far as to test American and Egyptian mummies, establishing the ancient nature of the ABO polymorphisms. In 1954 Mourant drew together the rapidly expanding body of blood group data in the first comprehensive summary of human biochemical diversity, *The Distribution of the Human Blood Groups* – a seminal work that became the standard text of experimental human population genetics for the next twenty years. This was the beginning of the modern era of human genetics.

While the Hirszfelds clearly felt that their data on blood groups supported a racial classification that had become blurred by recent migration, and Carleton Coon later used them to support his theories of discrete subspecies, no one had actually tested the genetic data to see if there was any real indication of racial subdivision. This obvious analysis was finally carried out in 1972 by a geneticist whose primary research interest, oddly enough, was fruit flies – not humans.

Using the data collected by Mourant and others, Richard Lewontin, then a professor at the University of Chicago, performed a seemingly trivial study of how human genetic variation sorted into within- versus between-group components. The question he was tying to answer, objectively, was whether there was any indication in the genetic data of a distinct subdivision between human races. In other words, he was directly testing the hypotheses of Linnaeus and Coon about human subspecies. If human races showed significant differences in their patterns of genetic diversity, then Linnaeus and Coon must be right.

Lewontin describes the development of the analysis:

The paper was written in response to a request . . . to contribute an article to the new journal *Evolutionary Biology*. I had been thinking at that time about diversity measures . . . not in the context of population genetics, but in the context of ecology. I had to take a very long bus trip to Bloomington, Indiana, and I had long had the habit, when going on trains and buses, of writing papers. I needed to write this paper, so I went on the bus trip with a copy of Mourant and a table of p*ln*p [a mathematical table used for calculating the diversity measure].

On this bus trip, he began what would become one of the landmark studies in human genetics. In the analysis, Lewontin used as his model the new science of biogeography (the study of animal and plant geographic distributions) because he thought this was analogous to what he was doing with humans – looking for geographic subdivisions in order to define race. In fact, unsure of how to define a 'race' objectively, he divided humans largely along geographical lines – Caucasians (western Eurasia), Black Africans (sub-Saharan Africa), Mongoloids (east Asia), South Asian Aborigines (southern India), Amerinds (Americas), Oceanians and Australian Aborigines.

The surprising result he obtained was that the majority of the genetic differences in humans were found within populations – around 85 per cent of the total. A further 7 per cent served to differentiate populations within a 'race', such as the Greeks from the Swedes. Only 8 per cent were found to differentiate between human races. A startling conclusion – and clear evidence that the subspecies classification should be scrapped. Lewontin says of the result:

I had no expectation – I honestly didn't. If I had any prejudice, it probably was that the between-race difference would have been a lot larger. This was reinforced by the fact that, when my wife and I were in Luxor [Egypt], years before it was overrun with tourists, she got in a discussion with a guy in the lobby. He was talking to her as if he knew her. She kept saying 'I'm sorry, sir, you've mistaken me for someone else.' Finally he said, 'Oh, I'm sorry madam – you all look alike to me.' That really had a big effect on my thinking – they really are different from us, and we're all alike.

But the result was there in the statistical analysis, and it has been confirmed by many other studies over the past three decades. The small proportion of the genetic variation that distinguishes *between* human populations has been debated endlessly (is it higher within or between races?), but the fact remains that a small population of humans still retains around 85 per cent of the total genetic diversity found in our species. Lewontin likes to give the example that if a nuclear war were to happen, and only the Kikuyu of Kenya (or the Tamils, or the Balinese . . .) survived, then that group would still have

85 per cent of the genetic variation found in the species as a whole. A strong argument indeed against 'scientific' theories of racism – and clear support for Darwin's assessment of human diversity in the 1830s. It really was a case of 'out of many, one', as the title of this chapter says in Latin. But does this mean that the study of human groups is meaningless – can genetics really tell us anything about human diversity?

Forcing the issue

For the next step on our journey, we need to cover some basic population genetics. The theory of how genes in a population behave over time is fairly complicated, and makes use of many related branches of quantitative science. Statistical mechanics, probability theory and biogeography have all contributed to our understanding of population genetics. But many of the theoretical frameworks are based on a few key concepts that can be understood by anyone, reflecting the relative simplicity of the forces involved.

The most basic force is mutation, and without it polymorphism would not exist. By mutation I mean a random change in a DNA sequence – these occur at a rate of around thirty per genome per generation. Looking at it another way, each person alive today is carrying around thirty completely novel mutations that distinguish them from their parents. Mutations are random because they arise as copying mistakes during the process of cell division, with no particular rhyme or reason as to where those mistakes might occur – our genomes do not appear to favour certain types of mutation based on what the effect might be. Rather, we are like Heath Robinson engineers, forced to make use of what we are given in the mutational lottery. The blood group variants discovered by Landsteiner originated as mutations, as do all other polymorphisms.

The second force is known as selection, in particular natural selection. This is the force that Darwin got so excited about, and it has certainly played a critical role in the evolution of *Homo sapiens*. Selection acts by favouring certain traits over others by conferring a reproductive advantage on their bearers. For example, in cold climates animals with thick fur would have an advantage over hairless ones, and their offspring would be more likely to survive. Selection is certainly what has made us the sentient, cultured apes we are today. It is what produced the important traits of speech, bipedalism and opposable thumbs. Without natural selection we would still be very similar to the relatively unsophisticated ape-like ancestor we would encounter if we could go back in time 5 million years or so.

The third force is known as genetic drift. This is a rather specialized term for something we have an innate sense of – the tendency of small samples to reflect a biased view of the population from which they are drawn. If you flip a coin 1,000 times, you expect to get around 500 heads and 500 tails. If, on the other hand, you flip the coin only 10 times, it is quite likely that you will get something other than a 5–5 outcome – perhaps 4–6 or 7–3. This random fluctuation in a sampled group is due to the small number of individual events in the sample. If we think of people as genetically sampled 'events', and assume that the population from which we will draw the sample for the *next* generation is created anew in the *present* generation (as is the case for living organisms), then you can see that small population sizes can lead to drastic changes in gene frequency within only a few generations. In the case of our coin flippers, getting a result of 7–3 would be reflected in the likelihood of flipping that number in the next generation, with a 70 per cent chance of getting heads and 30 per cent of getting tails. It's like a ratchet, because the probability change in the previous generation affects the probability in the subsequent generations. In the coin-flipping analogy, we've gone from a frequency of 50 per cent to 70 per cent in a single generation – a pretty rapid change. Clearly, drift can have a huge effect on gene frequencies in small populations.

The combination of these three forces has produced the dizzying array of genetic patterns we see today – and the vast diversity we see in human populations. Their action has also produced the small percentage of human variation that distinguishes between human groups. That much was known by the middle of the twentieth century. But simply recognizing the existence of human diversity at a biochemical level, and knowing something about the way genes behave in populations, didn't really say much about the details of human evolution and migration. Enter an Italian physician with a historical bent and a talent for mathematics, who came to the field influenced by a new way of thinking about bacteria and flies.

The Italian job

Luigi Luca Cavalli-Sforza had started his career in Pavia as a medical student. He soon left medicine to devote himself to genetics research, first on bacteria and later on humans. At university he had studied under the famous *Drosophila* geneticist Buzzati-Traverso, who was an adherent of the Dobzhansky school of genetics. Theodosius Dobzhansky had also been Richard Lewontin's PhD supervisor, and the story therefore begins to show a common thread. The main theme of Dobzhansky's research was the study of genetic variation, in particular large-scale chromosomal rearrangements in fruit flies. He pioneered techniques in genetic analysis, and his laboratory in New York was to be the epicentre of a revolution in biology during the mid-twentieth century. Dobzhansky and his students advocated a new view of genetic variation in which there was no division into an optimized 'wild type' (the normal form of the organism, created through a long period of natural selection) and a quirky 'mutant', invariably disadvantaged in some way. This was too simplistic, they thought – primarily because there was simply too much variation to account for if most of the mutants were carrying a suboptimal genetic package. If, instead, one thought of variation as the normal state of species, then evolution suddenly made much more sense. There was a previously unrecognized reservoir of different types on which evolution could act – favouring some in one case, losing them in another.

So, with a thorough background in the seemingly disparate fields of fruit-fly variation and medicine, Cavalli-Sforza began to conduct studies of blood polymorphisms – later termed 'classical' polymorphisms by geneticists – in an effort to assess the relationships among modern humans. This work was begun in the 1950s, a heady time for the field of genetics. The structure of DNA had just been deciphered by Crick and Watson, and the application of the methodology of the physical sciences promised a revolution in biology. Like most geneticists, Cavalli-Sforza made use of the rapidly developing techniques of biochemistry to assay variation. But unlike many of them, he was also comfortable with the application of mathematics – particularly its most pragmatic branch, statistics. The dizzying variety of the data being generated by studies of polymorphisms needed a coherent theoretical framework to make it understandable. And statistics was about to ride to the rescue.

Imagine a group of anything that exhibits variation – the different colours of stones in a streambed, snail-shell size, fruit-fly wing length, or human blood groups. At first glance these variations seem random and disconnected. If we have multiple sets of such objects, then it seems more complex still – even chaotic. What does it reveal about the mechanism by which the diversity was generated?

The knee-jerk reaction of most biologists in the 1950s to any pattern of diversity in nature was that selection was the root cause. Human diversity was no exception, as the eugenicists made quite clear. In part this stemmed from the widespread belief in 'wild types' and 'mutants'. The wild type could encompass any trait – size, colour, nose shape, or any other 'normal' characteristic of the organism. This was reinforced by the fact that genetic diseases (which were clearly 'abnormal') were some of the first variants recognized in humans, setting the stage for a worldview in which people were categorized as fit or unfit according to a Darwinian evolutionary struggle. However, in the 1950s Motoo Kimura, a Japanese scientist working in the United States, began to do some genetic calculations using methods originally derived for analysing the diffusion of gases, formalizing work carried out by Cavalli-Sforza and others. This work would eventually lead the field out of the 'mutant' morass.

Kimura noticed that genetic polymorphisms in populations can vary in frequency owing to random sampling errors – the 'drift' mentioned above. What was exciting in his results was that drift seemed to change gene frequencies at a predictable rate. The difficulty with studying selection was that the *speed* with which it produced evolutionary change depended entirely on the *strength* of selection – if the genetic variant was extremely fit, then it increased in frequency rapidly. However, it was virtually impossible to measure the strength of selection experimentally, so no one could make predictions about the rate of change. In our coin-flipping example, if heads is one variant of a gene and tails is another, then the increase in frequency from 50 per cent to 70 per cent in a single 'generation' would imply very strong selection favouring *heads*. Clearly, though, this isn't the case – heads increased to 70 per cent for reasons that had nothing to do with how well adapted it was.

Kimura's insight was that most polymorphisms appear to behave like this – that is they are effectively free from selection, and thus they can be treated as evolutionarily 'neutral', free to drift around in frequency due entirely to sampling error. There has been great debate among biologists about the fraction of polymorphisms that are neutral – Kimura and his scientific followers thought that almost all genetic variation was free from selection, while many scientists continue to favour a significant role for natural selection. Most of the polymorphisms studied by human geneticists, though, had probably arrived at their current frequencies because of drift. This opened the door to a new way of analysing the rapidly accumulating data on blood group polymorphisms. But before that could happen, the field needed to make a quick detour through the Middle Ages.

'Ock the Knife'

William of Ockham (*c.*1285–1349) was a medieval scholar who must have been a nightmare to be around. Ockham believed literally in Aristotle's statement that 'God and nature never operate superfluously, but always with the least effort', and took every opportunity to invoke his interpretation of this view in arguments with his colleagues. Ockham's razor, as it became known, was stated quite simply in Latin: *Pluralitas non est ponenda sine necessitate* (plurality is not to be posited without necessity). In its most basic form, Ockham's statement is a philosophical commitment to a particular view of the universe – a view that has become known as parsimony. In the real world, if each event occurs with a particular probability, then multiple events occur with multiplied probabilities and, overall, the complex events are less likely than the simple ones. It is a way of breaking down the complexity of the world into understandable parts, favouring the simple over the absurd. I may actually fly from Miami to New York via Shanghai – but it is not terribly likely.

This may seem trivial when applied to my travel schedule, but it is not so obvious when we start to apply it to the murky world of science. How do we really know that nature always takes the most parsimonious path? In particular, is it self-evident that 'simplify' is nature's buzzword? This book is not the forum for a detailed discussion of the history of parsimony (there are several references in the bibliography where the subject is discussed in great detail), but it seems that nature usually does favour simplicity over complexity. This is particularly true when things change – like when a stone drops from a cliff to the valley below. Gravity clearly exerts itself in such a way that the stone moves directly – and rather quickly – from the high to the low point, without stopping for tea in China.

So, if we accept that when nature changes, it tends to do so via the shortest path from point A to point B, then we have a theory for inferring things about the past. This is quite a leap, since it implies that by looking at the present we can say something about what happened before. In effect, it provides us with a philosophical time machine with which to travel back and dig around in a vanished age. Pretty impressive stuff. Even Darwin was an early adherent – Huxley actually scolded him on one occasion for being such a stick-in-the-mud about his belief that *natura non facit saltum* (nature doesn't make leaps).

The first application of parsimony to human classification was published by Luca Cavalli-Sforza and Anthony Edwards in 1964.* In this study they made two landmark assumptions which would be adopted in each subsequent study of human genetic diversity. The first was that the genetic polymorphisms were behaving as Kimura had predicted – in other words, they were all neutral, and thus any differences in frequency were due to genetic drift. The second assumption was that the correct relationship among the populations must adhere to Ockham's rule, minimizing the amount of change required to explain the data. With these key insights, they derived the first family tree of human groups based on what they called the 'minimum evolution' method. In effect, this means that the populations are linked in a diagram such that the ones with the most similar gene frequencies are

---

* Parsimony here is simply the application of methods which infer evolutionary history in such a way as to minimize complexity. It is not necessarily the method known as 'maximum parsimony' used by many population geneticists.

closest together, and that overall the relationship among the groups minimizes the total magnitude of gene frequency differences.

Cavalli-Sforza and Edwards looked at blood group frequencies from fifteen populations living around the world. The result of this analysis, laboriously calculated by an early Olivetti computer, was that Africans were the most distant of the populations examined, and that European and Asian populations clustered together. It was a startlingly clear insight into our species' evolutionary history. As Cavalli-Sforza says modestly, the analysis 'made some kind of sense', based on their concept of how human populations should be related – European populations were closer to each other than they were to Africans, New Guineans and Australians grouped together, and so on. This was a reflection of similarities in gene frequencies, and since these frequencies changed in a regular way over time (remember genetic drift), it meant that the time elapsed since Europeans started diverging from each other was less than the time separating Europeans from Africans. The old monk had proven useful after 700 years – and anthropology had a way forward.*

With this new approach to human classification, it was even possible to calculate the dates of population splits, making several assumptions about the way humans had behaved in the past, and the sizes of the groups they lived in. This was first done by Cavalli-Sforza and his colleague Walter Bodmer in 1971, yielding an estimate of 41,000 years for the divergence between Africans and East Asians, 33,000 for Africans and Europeans and 21,000 for Europeans and East Asians. The problem was, it was uncertain how reasonable their assumptions about population structure really were. And crucially, it still failed to provide a clear answer to the question of where humans had originated. What the field needed now was a new kind of data.

---

* Cavalli-Sforza and Edwards also developed other methods of analysing the relationship between populations on the basis of gene frequencies which rely less on absolute minimization of evolutionary change. Parsimony, however, is still widely used in the field.

Alphabet soup

Emile Zuckerkandl was a German-Jewish émigré working at the California Institute of Technology in Pasadena. He spent much of his scientific career tenaciously focused on one problem: the structure of proteins. Working with the Nobel Prize-winning biochemist Linus Pauling in the 1950s and 60s, Zuckerkandl studied the basic structure of the oxygen-carrying molecule haemoglobin – chosen because it was plentiful and easy to purify. Haemoglobin had another important characteristic: it was found in the blood of every living mammal.

Proteins are composed of a linear sequence of amino acids, small molecular building blocks that combine in a unique way to form a particular protein. The amazing thing about proteins is that, although they do their work twisted into baroque shapes, often with several other proteins sticking to them in a complex way, the ultimate form and function of the active protein is determined by a simple linear combination of amino acids. There are twenty amino acids used to make proteins, with names like lysine and tryptophan. These are abbreviated by chemists to single letter codes – K and Y in this case.

Zuckerkandl noticed an interesting pattern in these amino acid sequences. As he started to decipher haemoglobins from different animals, he found that they were similar. Often they had identical sequences for ten, twenty, or even thirty amino acids in a row, and then there would be a difference between them. What was fascinating was that the more closely related the animals were, the more similar they were. Humans and gorillas had virtually identical haemoglobin sequences, differing only in two places, while humans and horses differed by fifteen amino acids. What this suggested to Zuckerkandl and Pauling was that molecules could serve as a sort of molecular clock, documenting the time that has elapsed since a common ancestor through the number of amino acid changes. In a paper published in 1965, they actually refer to molecules as 'documents of evolutionary history'. In effect, we all have a history book written in our genes. According to Zuckerkandl and Pauling, the pattern written in a molecular structure can even provide us with a glimpse of the ancestor

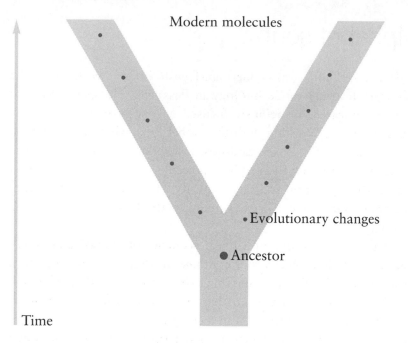

Figure 1 The evolutionary 'genealogy' of two related molecules, showing sequence changes accumulating on each lineage.

itself, making use of Ockham's razor to minimize the number of inferred amino acid changes and working back to the likely starting point (see Figure 1). Molecules are, in effect, time capsules left in our genomes by our ancestors. All we have to do is learn to read them.

Of course, Zuckerkandl and Pauling realized that proteins were not the ultimate source of genetic variation. This honour lay with DNA, the molecule that actually forms our genes. If DNA encodes proteins (which it does), then the best molecule to study would be the DNA itself. The problem was that DNA was extremely difficult to work with, and getting a sequence took a long time. In the mid-1970s, however, Walter Gilbert and Fred Sanger independently developed methods for rapidly obtaining DNA sequences, for which they shared the Nobel Prize in 1977. The ability to sequence DNA set off a revolution in biology that has continued to this day, culminating in 2000 with the completion of a working draft of the entire human genome sequence. DNA research has revolutionized the way we think about biology, so it isn't surprising that it has had a significant effect on anthropology as well.

The crowded garden

So we find ourselves in the 1980s with the newly developed tools of molecular biology at our disposal, a theory for how polymorphisms behave in populations, a way to estimate dates from molecular sequence data and the burning question of how genetics can answer a few age-old questions about human origins. What the field needed now was a lucky insight and a bit of chutzpah. Both of these were to be found in the early 1980s in the San Francisco Bay area of northern California.

Allan Wilson was an Australian biochemist working at the University of California, Berkeley, on methods of evolutionary analysis using molecular biology – the new branch of biology that focused on DNA and proteins. Using the methods of Zuckerkandl and Pauling, he and his students had used molecular techniques to estimate the date of the split between humans and apes, and they had also deciphered some of the intricate details of how natural selection can tailor proteins to their environments. Wilson was an innovative thinker, and he embraced the techniques of molecular biology with a passion.

One of the problems that molecular biologists encountered in studying DNA sequences was that of the duplicate nature of the information. Inside each of our cells, what we think of as our genome – the complete DNA sequence that encodes all of the proteins made in our bodies, in addition to a lot of other DNA that has no known function – is really present in two copies. The DNA is packaged into neat, linear components known as chromosomes – we have twenty-three pairs of them. Chromosomes are found inside a cellular structure known as the nucleus. One of the main features of our genome is the astounding compartmentalization – like computer folders within folders within folders. In all there are 3,000,000,000 (3 billion) building blocks, known as nucleotides (which come in four flavours: A, C, G and T), in the human genome, and we need some way to get at all of the information it contains in a straightforward way. This is why we have chromosomes, and why they are kept squirrelled away from the rest of the cell inside the nucleus.

The reason we have two copies of each chromosome is more complicated, but it comes down to sex. When a sperm fertilizes an egg, one of the main things that happens is that part of the father's genome and part of the mother's genome combine in a 50 : 50 ratio to form the new genome of the baby. Biologically speaking, one of the reasons for sex is that it generates new genomes every generation. The new combinations arise, not only at the moment of conception with the 50 : 50 mixing of the maternal and paternal genomes, but also prior to that, when the sperm and egg themselves are being formed. This pre-sexual mixing, known as genetic recombination, is possible because of the linear nature of the chromosomes – it is relatively easy to break both chromosomes in the middle and reattach them to their partners, forming new, chimeric chromosomes in the process. The reason why this occurs, as with the mixing of Mum's and Dad's DNA, is that it is probably a good thing, evolutionarily speaking, to generate diversity in each generation. If the environment changes, you'll be ready to react.

But wait, you might say, why are these broken and reattached chromosomes any different from the ones that existed before? They were supposed to be duplicates! The reason, quite simply, is that they aren't exact copies of each other – they differ from each other at many locations along their length. They are like duplicates of duplicates of duplicates of duplicates, made with a dodgy copying machine that introduces a small number of random errors every time the chromosomes are copied. These errors are the mutations mentioned above, and the differences between each chromosome in a pair are the polymorphisms. Polymorphisms are found roughly every 1,000 nucleotides along the chromosome, and serve to distinguish the chromosomes from each other. So, when recombination occurs, the new chromosomes are different from the parental types.

The evolutionary effect of recombination is to break up sets of polymorphisms that are linked together on the same piece of DNA. Again, this diversity-generating mechanism is a good thing evolutionarily speaking, but it makes life very difficult for molecular biologists who want to read the history book in the human genome.

Recombination allows each polymorphism on a chromosome to behave independently from the others. Over time the polymorphisms are recombined many, many times, and after hundreds or thousands of generations, the pattern of polymorphisms that existed in the common ancestor of the chromosomes has been entirely lost. The descendant chromosomes have been completely shuffled, and no trace of the original deck remains. The reason this is bad for evolutionary studies is that, without being able to say something about the ancestor, we cannot apply Ockham's razor to the pattern of polymorphisms, and we therefore have no idea how many changes really distinguish the shuffled chromosomes. At the moment, all of our estimates of molecular clocks are based on the rate at which new polymorphisms appear through mutation. Recombination makes it look like there have been mutations when there haven't, and because of this it causes us to overestimate the time that has elapsed since the common ancestor.

One of the insights that Wilson and several other geneticists had in the early 1980s was that if we looked outside of the genome, at a small structure found elsewhere in the cell known as the mitochondrion, we might have a way of cheating the shuffle. Interestingly, the mitochondrion has its own genome – it is the only cellular structure other than the nucleus that does. This is because it is actually an evolutionary remnant from the days of the first complex cells, billions of years ago – the mitochondrion is what remains of an ancient bacterium which was swallowed by one of our single-celled ancestors. It later proved useful for generating energy inside the cell, and now serves as a stream-lined sub-cellular power plant, albeit one that started life as a parasite. Fortunately, the mitochondrial genome is present in only one copy (like a bacterial genome), which means that it can't recombine. Bingo. It also turns out that, instead of having one polymorphism roughly every 1,000 nucleotides, it has one every 100 or so. To make evolutionary comparisons we want to have as many polymorphisms as possible, since each polymorphism increases our ability to distinguish between individuals. Think of it this way: if we were to look at only one polymorphism, with two different forms A and B, we would sort everyone into two groups, defined only by variant A or variant B. On the other hand, if we looked at ten polymorphisms with two variants each, we would have much better resolution, since the likelihood of multiple individuals having exactly the same set of variants is much lower. In other words, the more polymorphisms we have, the better our chances of inferring a useful pattern of relationships among the people in the study. Since polymorphisms in mitochondrial DNA (mtDNA) are ten times more common than in the rest of our genome, it was a good place to look.

Rebecca Cann, as part of her PhD work in Wilson's laboratory, began to study the pattern of mtDNA variation in humans from around the world. The Berkeley group went to great lengths to collect samples of human placentas (an abundant source of mtDNA) from many different populations – Europeans, New Guineans, Native Americans and so on. The goal was to assess the pattern of variation for the entire human species, with the aim of inferring something about human origins. What they found was extraordinary.

Cann and her colleagues published their initial study of human mitochondrial diversity in 1987. It was the first time that human DNA polymorphism data had been analysed using parsimony methods to infer a common ancestor and estimate a date. In the abstract to the paper they state the main finding clearly and succinctly: 'All these mitochondrial DNAs stem from one woman who is postulated to have lived about 200,000 years ago, probably in Africa.' The discovery was big news, and this woman became known in the tabloids as Mitochondrial Eve – the mother of us all. In a rather surprising twist, though, she wasn't the only Eve in the garden – only the luckiest.

The analysis performed by Cann and her colleagues involved asking how the mtDNA sequences were related to each other. In their paper they assumed that if two mtDNA sequences shared a sequence variant at a polymorphic site (say, a C at a position where the sequences had either a C or a T), then they shared a common ancestor. By building up a network of the mtDNA sequences – 147 in all – they were able to infer the relationships between the individuals who had donated the samples. It was a tedious process, and involved a significant amount of time analysing the data on a computer. What their results showed were that the greatest divergence between mtDNA sequences was actually found among the Africans – showing that they had been diverging for longer. In other words, Africans are the oldest group on the planet – meaning that our species had originated there.

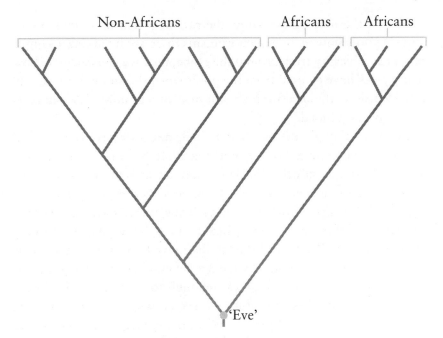

Figure 2 Proof that modern humans originated in Africa – the deepest split in the genealogy of mtDNA ('Eve') is between mtDNA sequences from Africans, showing that they have been accumulating evolutionary changes for longer.

One of the features of the parsimony analysis used by Cann, Stoneking and Wilson to analyse their mtDNA sequence data is that it inevitably leads back to a single common ancestor at some point in the past. For any region of the genome that does not recombine – in this case, the mitochondrion – we can define a single ancestral mitochondrion from which all present-day mitochondria are descended. It is like looking at an expanding circle of ripples in a pond and inferring where the stone must have dropped – in the dead centre of the circle. The evolving mtDNA sequences, accumulating polymorphisms as they are passed from mother to daughter, are the expanding waves, and the ancestor is the point where the stone entered the water. By applying Zuckerkandl and Pauling's methods of analysis, we can 'see' the single ancestor that lived thousands of years ago, and which has mutated over time to produce all of the diverse forms that exist today. Furthermore, if we know the rate at which mutations occur, and we know how many polymorphisms there are by taking a sample of human diversity from around the globe, then we can calculate how many years have elapsed from the point when the stone dropped – in other words, to the ancestor from whom all of the mutated descendants must have descended.

Crucially, though, the fact that a single ancestor gave rise to all of the diversity present today does not mean that this was the only person alive at the time – only that the descendant lineages of the other people alive at the same time died out. Imagine a Provençal village in the eighteenth century, with ten families living there. Each has its own special recipe for bouillabaisse, but it can only be passed on orally from mother to daughter. If the family has only sons, then the recipe is lost. Over time, we gradually reduce the number of starting recipes, because some families aren't lucky enough to have had girls. By the time we reach the present century we are left with only one surviving recipe – *la bouillabaisse profonde*. Why did this one survive? By chance – the other families simply didn't have girls at some point in the past, and their recipes blew away with the *mistral*. Looking at the village today, we might be a little disappointed at its lack of culinary diversity. How can they all eat the same fish soup?

Of course, in the real world, no one transmits a recipe from one generation to the next without modifying it slightly to fit her own tastes. An extra clove of garlic here, a bit more thyme there, and *voilà*! – a bespoke variation on the *matrimoine*. Over time, these variations on a theme will produce their own diversity in the soup bowls – but the recipe extinction continues none the less. If we look at the bespoke village today we see a remarkable diversity of recipes – but they can *still* be traced back to a single common ancestor in the eighteenth century, thanks to Ock the Knife. This is the secret of Mitochondrial Eve.

The results from the 1987 study by Cann and her colleagues were followed up by a more detailed analysis a few years later, and both studies pointed out two important facts: that human mitochondrial diversity had been generated within the past 200,000 years, and that the stone had dropped in Africa. So, in a very short period of time – at least in evolutionary terms – humans had spread out of Africa to populate the rest of the world. There were some technical objections to the statistical analysis in the papers, but more extensive recent studies of mitochondrial DNA have confirmed and extended the conclusions of the original analysis. We all have an African great-great . . . grandmother who lived approximately 150,000 years ago.

Darwin, in his 1871 book on human evolution *The Descent of Man, and Selection in Relation to Sex*, had noted that 'in each great region of the world the living mammals are closely related to the extinct species of the same region. It is therefore probable that Africa was formerly inhabited by extinct apes closely allied to the gorilla and chimpanzee; and as these two species are now man's nearest allies, it is somewhat more probable that our early progenitors lived on the African continent than elsewhere.' In some ways this statement is incredibly far-sighted, since most nineteenth-century Europeans would have placed Adam and Eve in Europe or Asia. In other ways it is rather trivial, since apes originated in Africa around 23 million years ago, so if we go back far enough we are eventually bound to encounter our ancestors in this continent. The key is to give a date – and this is why the genetic results were so revolutionary.

Anthropologists such as Carleton Coon had argued for the origin of human races through a process of separate speciation events from ape-like ancestors in many parts of the world. This hypothesis became known as multiregionalism, and it persists in some anthropological circles even today. The basic idea is that ancient hominid, or human-like, species migrated out of Africa over the course of the past couple of million years or so, establishing themselves in east Asia very early on, and then evolving *in situ* into modern-day humans – in the process creating the races identified by Coon. To understand why this theory was so widely accepted, we need to leave aside DNA for a while and rummage around in some old bones.

Dutch courage

Linnaeus named our species *Homo sapiens*, Latin for 'wise man', because of our uniquely well-developed intellect. Since the nineteenth century, however, it has been known that other hominid species existed in the past. In 1856, for instance, a skull was discovered in the Neander Valley of western Germany. In pre-Darwinian Europe the bones were originally thought to be the remains of a malformed modern human, but it was later found to be a widespread and distinct species of ancestral hominid, christened Neanderthal Man after the site of its discovery. This was the first scientific recognition of a human ancestor, and provided concrete evidence that the hominid lineage has evolved over time. By the end of the nineteenth century the race was well and truly on to find other 'missing links' between humans and apes. And in 1890 a doctor working for the Dutch East India company in Java hit the jackpot.

Eugène Dubois was obsessed with human evolution, and his medical appointment in the Far East was actually part of an elaborate plan to bring him closer to what he saw as the cradle of humanity. Born in 1858 in Eijsden, Holland, Dubois specialized in anatomy at medical school. By 1881, he had been appointed as an assistant at the University of Amsterdam, but he found academic life to be too confining and hierarchical. So, in 1887, he packed up his worldly belongings and convinced his wife to set off with him on a quest to find hominid remains.

Dubois believed that humans were most closely related to gibbons, a species of ape only found in the Indo-Malaysian archipelago. This was because of their skull morphology (lack of a massive, bony crest on the top and a flatter face than that found in other apes) and the fact that they sometimes walked erect on their hind legs – both reasonable enough pieces of evidence, he thought, to look for the missing link in south-east Asia. His first excavations in Sumatra yielded only the relatively recent remains of modern humans, orang-utans and gibbons, but when he turned his attention to Java his luck changed.

In 1890 Dubois was sifting through fossils recovered from a river-bank at Trinil, in central Java, when he found a rather odd skullcap. To him it looked like the remains of an extinct chimpanzee known as *Anthropopithecus*, although without benefit of a good anatomical collection for comparison (he was in a colonial outpost, after all) it was difficult to be certain. The following year, however, a femur recovered from the same location threw the specimen into a whole new light. The leg bone was clearly not from a climbing chimpanzee, but rather from a species that walked upright. His calculations of the cranial capacity, or brain size, of the new find, in combination with its upright stance, led him to make a bold leap of faith. He named the new species *Pithecanthropus erectus*, Latin for 'erect ape-man'. This was the missing link everyone had been searching for.

The main objection to Dubois' discovery – battled out in public debates and carefully worded publications over the next decade – was that there was very little evidence that the skull and femur (and a tooth that was later found at the site) had actually come from the same individual. They were excavated at different times, and the relationship between the soil layers from which they had been recovered was unknown. Later finds of *Pithecanthropus* did reveal the Trinil femur to be anomalous, and it seems likely that it actually belongs to a more modern human. The tooth may well be that of an ape. Despite this, and despite Dubois' incorrect assertion that the remains proved that modern humans had originated in south-east Asia from gibbon-like ancestors, the discovery of the Trinil skullcap was a watershed event in anthropology. The Javanese ape-man was clearly a long-extinct human ancestor – one with a cranial capacity much lower than our own, but still far above the range seen in apes. Although he got it wrong in so many ways, Dubois had got it right where it counted.

The competition to find other hominid remains intensified in the early twentieth century, with the lion's share of the activity focused on east Asia and Africa. The discovery of *Pithecanthropus*-like fossils in the 1920s and 30s at Zhoukoudian, China, showed that Dubois' ape-man had been widespread in Asia. The uniting of the Zhoukoudian *Sinanthropus* ('Peking man') with *Pithecanthropus* ('Java man') in the 1950s provided the first clear evidence for a widespread, extinct species of hominid: *Homo erectus*. But the most amazing finds were to come from Africa, starting with the work of Raymond Dart in the 1920s.

In 1922 Dart was appointed Professor of Anatomy at the University of the Witwatersrand in South Africa. This must have come as a bit of a blow to the academically high-flying Australian (who was previously based in Britain), since 'Wits' at that time was a scientific backwater. Nonetheless, he set about building the foundation of an academic Department of Anatomy in the newly created university, which involved the establishment of a collection of anatomical specimens.

He urged his students to send him material, and after one of them found a fossil baboon skull from a quarry at Taung, near Johannesburg, Dart felt that he was on to something interesting.

Up to this point, most fossilized human remains had come from Europe and Asia: Neanderthal, Peking Man, Java Man – all were found outside Africa. In 1921, however, a Neanderthal-like skull was unearthed in Northern Rhodesia (now Zambia), proving that Africa had an ancient hominid pedigree as well. Dart was well aware of this when he contacted the owner of the Taung quarry to send him additional samples of material. What he found in the first crates to arrive in the summer of 1924 was, to his great delight, the oldest human fossil yet discovered.

As he painstakingly picked off the compressed rubbish accumulated over aeons in the Taung cave, Dart revealed an ape-like face staring back at him. Its small size and intact milk teeth immediately gave it away as a child's skull, and Dart's estimate of its cranial capacity revealed it to be well within the normal range found in modern apes – around 500 cubic centimetres. What was odd about the find was the size of the canine teeth – much smaller than those of apes – and the location of the foramen magnum, which serves as a conduit for the spinal column in its connection to the brain: it was orientated downward in the fossil, like modern humans, rather than backward, as is the case in apes. To Dart, both of these features indicated that the Taung baby, as it became known, was no ordinary simian. In a 1925 paper he asserted that the skull represented the remains of a new species, which he called *Australopithecus africanus* ('African southern ape'), that walked upright and used tools. In Dart's own words, the Southern Ape was 'one of the most significant finds ever made in the history of anthropology'. It was the first clear evidence for a missing link between apes and humans in Africa, and it set off a tidal wave of human-origins research that was to culminate a few decades later in universal acceptance for the African origin of humanity. However, most of this work was to focus on a region a few thousand miles away, in eastern Africa.

The African Rift Valley is part of a massive line of intense geological upheaval formed by the action of great tectonic plates that make up the earth's crust. Roughly 2,000 miles long, it stretches from Eritrea in the north to Mozambique in the south, and is most recognizable by the series of lakes along its length – Turkana, Victoria, Tanganyika and Malawi, among others. This longitudinal gash has been a cauldron of activity over the past 20 million years, with volcanoes, lakes, mountains and rivers coming and going at a brisk pace. For this reason, the accumulated layers of millions of years – soil, volcanic ash, lake sediments – are constantly being tossed about and exposed. When this happens in east Africa, interesting things often turn up – all you have to do is look for them.

Louis Leakey had grown up in Kenya. The son of English missionaries, and raised in a Kikuyu village, he had spent his life looking for fossil remains in the valleys and riverbeds of the Rift. In 1959 at Olduvai, in northern Tanzania, his search was to pay off. It was nearing the end of the field season and, with research funds running on empty, Louis and his wife Mary were preparing to return to Nairobi. On the way back to camp one evening Mary stumbled upon a skull exposed by a recent rockslide. After painstakingly excavating the fossil over the next three weeks, the Leakeys returned to their laboratory at the Kenyan National Museum. The detailed analysis of the remains revealed it to be an *Australopithecus*, the first to be found in east Africa. But the shocker came when the layer of sediment surrounding the skull was dated using the newly developed technique of isotopic analysis, which calculates age based on the rate of radioactive decay. The skull had been buried 1.75 million years ago. This nearly *doubled* the length of time that most scientists had allowed for human evolution. Yet here was a missing link, midway between apes and humans, dating from that time. The scientific world was amazed – and encouraged. The massive boost in funding that the Leakeys and their colleagues received in the wake of the Olduvai discovery enabled them to find many more Australopithecines in the Rift over the subsequent thirty years.

The discovery of the Southern Ape Man in east Africa pointed the way towards modern humans, but it was only when unequivocal members of our own genus, *Homo*, were discovered there in the 1960s and 70s that the African origin hypothesis became widely accepted. The earliest *Homo erectus* fossils yet discovered date from around 1.8 million years ago, and they were found in east Africa (the African variant of *Homo erectus* is sometimes given the name *Homo ergaster*). Recent discoveries in the medieval city of Dmanisi, in the former Soviet Republic of Georgia, show that they left Africa soon thereafter – perhaps reaching east Asia within 100,000 years. From this we can infer that all *Homo erectus* around the world last shared a common ancestor in Africa nearly 2 million years ago. But according to the Berkeley mitochondrial data, Eve lived in Africa less than 200,000 years ago. How can we reconcile the two results?

It's all about timing

Let's step back for a moment and consider the case objectively. The evidence for an African Genesis of *Homo erectus* is circumstantial – we see evolutionary 'missing links' in Africa, either uniquely or first. These include an unbroken chain of ancestral hominids stretching back more than 5 million years to the recently discovered chimpanzee-like apes *Ardipithecus*. But is this evidence sufficient to conclude that Africa was also the birthplace of our species? Perhaps, but fossils can be misleading. Imagine finding a perfectly preserved Neanderthal skeleton in south-western France, dated accurately to 40,000 years ago, and one of *Australopithecus*, in Africa, dated to 2 million years before. Of these two extinct hominids, separated in time by millions of years and in place by thousands of miles, which is actually more likely to be a direct ancestor of modern Europeans? Oddly enough, it is not the obvious choice. As we'll see later in the book, modern Europeans are almost certainly not the descendants of Neanderthals (despite what you may think of your colleague in the office next door), while the Southern Ape is, surprisingly, more likely to be our direct ancestor. Stones and bones inform our knowledge of the past, but they cannot tell us about our genealogy – only genes can do this.

So, the answer to our question about dates – how to reconcile 200,000 and 2 million – is that *Homo erectus*, despite its clear similarity to us, did not evolve into modern *Homo sapiens* independently in the far corners of the earth. Coon was wrong. Rather, the conclusion from the mitochondrial data is that modern humans evolved very recently in Africa, and subsequently spread to populate the rest of the globe, replacing our hominid cousins in the process. It's a ruthless business, and only the winners leave a genetic trace. Unfortunately, *Homo erectus* appears to have lost.

As we'll see, other genetic data corroborates the mitochondrial results, placing the root of the human family tree – our most recent common ancestor – in Africa within the past few hundred thousand years. Consistent with this result, all of the genetic data shows the greatest number of polymorphisms in Africa – there is simply far more variation in that continent than anywhere else. You are more likely to sample extremely divergent genetic lineages within a single African village than you are in whole of the rest of the world. The majority of the genetic polymorphisms found in our species are found uniquely in Africans – Europeans, Asians and Native Americans carry only a small sample of the extraordinary diversity that can be found in any African village.

Why does diversity indicate greater age? Thinking back to our hypothetical Provençal village, why do the bouillabaisse recipes change? Because in each generation, a daughter decides to modify her soup in a minor way. Over time, these small variations add up to an extraordinary amount of diversity in the village's kitchens. And – critically – the longer the village has been accumulating these changes, the more diverse it is. It is like a clock, ticking away in units of rosemary and thyme – the longer it has been ticking, the more differences we see. It is the same phenomenon Emile Zuckerkandl noted in his proteins – more time equals more change. So, when we see greater genetic diversity in a particular population, we can infer that the population is older – and this makes Africa the oldest of all.

But does the placement of the root of our family tree in Africa mean that Coon was right, and Africans are frozen in some sort of ancestral evolutionary limbo? Of course not – all of the branches on the family tree change at the same rate, both within and outside of Africa, so there are derived lineages on each continent. That is the reason we see greater diversity within Africa – each branch has continued to evolve, accumulating additional changes. One of the interesting corollaries of inferring a single common ancestor is that each descendant lineage continues to change at the same rate, and therefore all of the lineages are the same age. The time that has elapsed between my mitochondrial

DNA type and Eve's is exactly the same as that of an African cattle herder, or a Thai boat captain, or a Yanomami hunter from Brazil – we are all the recent descendants of a single woman who lived in Africa less than 150,000 years ago.

This result begs the question of where Eve actually lived – where in Africa was the Garden of Eden? In one sense this is a red herring, since we know that there were many women alive all over Africa at this time. But, phrasing the question slightly differently, we can ask which populations in Africa retain the clearest traces of our genetic ancestors. Although the diversity within Africa has not by any means been sampled exhaustively, the picture that has emerged is that the oldest genetic lineages are found in people living in eastern and southern Africa. What we can infer from this is that these populations have maintained a direct mitochondrial link back to Eve, while the rest of us have lost some of these genetic signals along the way. We'll pursue our search for Eden, using Adam as a guide, in the next chapter.

3
Eve's Mate

A woman without a man is like a fish without a bicycle.
Gloria Steinem

In the last chapter we met 'Eve' – the female ancestor of everyone alive today, who lived in Africa around 150,000 years ago. Based on the populations that seem to have retained the clearest genetic signals from our distant grandmother, we've begun our search for the location of the Garden of Eden. But before we go any further, we need to clarify Eve's uniqueness. She represents the root of the mitochondrial family tree, and as such she unites everyone around the world in a shared maternal history. However, it isn't necessarily the case that every part of our DNA should tell the same story. Because of sexual recombination, our genome is composed of a large number of blocks that have each evolved pretty much independently. Perhaps one region of DNA traces back to an origin in Indonesia, while another began its journey in Mexico. So is Eve's lineage unique in tracing a recent journey out of Africa?

The answer is that the rest of our genome shows essentially the same pattern as the mtDNA, although it tends to have a lower degree of resolution. Studies of polymorphisms in the beta-globin gene (which encodes the oxygen-carrying component of blood), the CD4 gene (which encodes a protein that helps to regulate the immune system) and a region of DNA on chromosome 21 all show that African populations are much more diverse than those living outside of Africa, and provide dates that are substantially less than 2 million years for the age of our common African ancestor. But the problem with using markers like these – from the 22 pairs of chromosomes that comprise the majority of our genome – is that the information tends to be shuffled over time. The further apart the polymorphisms are, the more likely it is that they have been shuffled. And because shuffling obscures the historical signal, this means that most of our genome isn't terribly useful for tracing migrations.

There is one piece of DNA, though, that has recently proven to be an invaluable tool for inferring details about human history – providing us with far greater resolution than we ever thought possible about the paths followed by our ancestors during their wanderings. It is the male equivalent of mtDNA, in that it is only passed from father to son. For this reason, it defines a uniquely male lineage – a counterpart to the female line illuminated by studying mtDNA. It is the *patrimoine* in our Provençal village, and the details of lineage extinction and diversification that went on with the soup recipes also apply to this piece of DNA. It is known as the Y-chromosome.

Now wait a minute, you might be saying – what's going on with all of this maternal and paternal lineage gibberish? I thought that the whole idea of sex was to mix the mother's and father's genomes in a 50 : 50 ratio to produce the child? Why do we have these oddities that break the rules? For the mitochondrial DNA the answer is easy – it is actually outside of what we think of as the human genome, an evolutionary remnant of a time when it was a parasitic bacterium living inside the earliest cells. The story for the Y is a bit more complicated.

One of the quirky features of sexual reproduction is that the chromosomes that actually determine our sex – the so-called sex chromosomes – are exceptions to the 50 : 50 sexual mixing rule. The double layout of our genomes, with two copies of each chromosome, fails us when we get to these chromosomes. This is because of the way in which sex is determined in most animals, through the presence of a mismatched sex chromosome. In the case of mammals, it is the male that is mismatched, with one X and one Y-chromosome. In females, the X-chromosome is present in two copies, like the other chromosomes, allowing normal recombination. In males, however, the Y only matches with the X in short regions at either end, which serve to align the sex chromosomes properly during cell division. The rest of the Y-chromosome, known as the non-recombining portion of the Y, is pretty much completely unrelated to the X. Thus it has no paired chromosome with which it can recombine, and so it doesn't. It is passed unshuffled from one generation to the next, for ever – exactly like the mitochondrial genome.

The Y turns out to provide population geneticists with the most useful tool available for studying human diversity. Part of the reason for this is that, unlike mtDNA, a molecule roughly 16,000 nucleotide units long, the Y is huge – around 50 million nucleotides. It therefore has many, many sites at which mutations may have occurred in the past. As we saw in the last chapter, more polymorphic sites give us better resolution – if we only had Landsteiner's blood types to work with, everyone would be sorted into four categories: A, B, AB and O. To put it another way, the landscape of possible polymorphisms is simply much larger for the Y. And critically, because of its lack of recombination, we are able to infer the order in which the mutations occurred on the Y – just like mtDNA. Without this feature, we can't use Zuckerkandl and Pauling's methods to define lineages, and Ock the Knife can't help us with the ancestors.

How does the Y manage to exist without recombination – doesn't this contradict the idea that we need to create diversity in case it's necessary to react to a changing environment? The short answer is that there almost certainly are negative evolutionary consequences to the lack of recombination – part of the reason for the low number of functional genes found on the Y. The number of active genes varies greatly among different parts of the genome. In the mitochondrion, for instance, there are thirty-seven. The total number of genes in the nuclear genome is around 30,000 – approximately 1,500 per chromosome, on average. Most of the thousands of genes that would have been found in the bacterial ancestor of the mitochondria have been lost over the past few hundred million years as mitochondria have become more parasitic, giving up autonomy for a cosseted life inside another cell. Some have actually been inserted into the nuclear DNA, leaving us in the odd situation of having small pieces of our genome that are bacterial in origin. So in the case of mitochondrial DNA, it does look like there was pressure for it to lose its genes, transferring the critical ones to the nucleus where recombination can keep them in shape for the evolutionary race.

We see the same pattern of gene loss for the Y-chromosome.

Although the average human chromosome has roughly 1,500 active genes, only twenty-one have been identified on the Y. Some of these are present in multiple, tandem copies – as though the copying machine stuttered as it was duplicating that gene at some point in the past; these are counted as a single gene in our tally. Interestingly, all of the twenty-one genes on the Y are involved in some way in the creation of 'maleness' – particularly the gene known as *SRY*, for 'Sex-determining *Region* of the *Y*', which is the master switch for creating a male out of an undifferentiated embryo. The rest have secondary functions involved in making men look (and act) like men. For the most part, though, the DNA that makes up the Y is devoid of any discernible function. It is so-called 'junk DNA', which means that it is transmitted from one generation to the next without conferring any utility. But while it may be biological junk, it is like gold dust to population geneticists.

As we have seen, we can only study human diversity by looking at differences – the language of population genetics is written in the polymorphisms that we all carry around with us. These differences define all of us as unique individuals – unless we have a twin, no other person in the world has an identical pattern of genetic polymorphisms. This is the insight behind a DNA 'fingerprint', used to identify criminals. Applied to the Y-chromosome, it allows us to trace a unique male lineage back in time, from son to father to grandfather, and so on. Taken to the extreme, it allows us to travel back in time from the DNA of any man alive today to our first male ancestor – Adam. But how does it link unrelated men to each other in regional patterns? Surely each man must trace his own unique Y-chromosome line back to Adam?

The answer is no, but the reason is a bit complicated. It's because we're not as unrelated as we think. Imagine the situation for the majority of our genome – the parts that don't come uniquely from our mother or our father. Since we inherit half of this DNA from each of our parents, the pattern of polymorphisms it contains can be used to infer paternity, since it connects us to both our mother and our father. If my DNA is shown in court to have a 50 per cent match with that of a child I've never met, it is likely that I will be paying for the support of that child for many years to come – the probability of a match occurring by chance is infinitesimally small. So polymorphisms define us, and our parents, as part of a unique genealogical branch. No other group of people on earth has exactly the same story written in its DNA.

If we extend this further, and begin to think about our grandparents, and their grandparents, and so on, we lose some of the signal in each generation. I have a 50 per cent match with my father, but only a 25 per cent match with my grandfather, and only a 6 per cent match with his grandfather. This is because we acquire new ancestors in each generation as we go back in time, and they start to pile up pretty quickly. Each of my parents had two parents, and each of them had two parents, and so on. The geneticist Kenneth Kidd, of Yale University, has pointed out that if we double the number of ancestors in each generation (around twenty-five years), when we go back in time about 500 years each of us must have had over a million living ancestors. If we go back to the time of the Norman invasion of England, around a thousand years, our calculation tells us that we must have had over one trillion (1,000,000,000,000) ancestors – far more than the total number of people that have existed in the whole of human history. So what's going on? Is our calculation flawed in some way?

The answer is yes and no. The maths is certainly correct – the power of exponential growth has been known since at least the time of the Greeks, and we're all acquainted with the real-world phenomenon of 'breeding like rabbits'. The error in our ancestor tally stems not from a malfunctioning calculator, but from the assumption that each of the people in our genealogy is completely unrelated to the others. Clearly, people must share quite a bit of their ancestry, or we can't make the numbers work. This would have the effect of multiplying by a number smaller than two in each generation – in fact, for most people the number is pretty close to one. And the reason for this can be found by doing a bit of poetic bird-watching.

Water, water everywhere . . .

Samuel Taylor Coleridge, Romantic poet, failed classicist and drug addict, spent 1797–8 living in a small Dorset village. In between vigorous walks in the hills and long discussions with his neighbour, William Wordsworth, Coleridge found time for a fit of literary activity that was to produce his two greatest pieces of work, *Kubla Khan* and *The Rime of the Ancient Mariner*. The former, composed subconsciously while in an opium-induced dream state – how better to conjure up the 'stately pleasure dome' – is an extraordinary exercise in literary imagery. The latter, written during a more sober period, follows the travails of a ship in the South Seas. The mariner in the poem callously violates one of the unwritten laws of the sea by killing an albatross, and the entire crew are made to suffer the consequences, ending up becalmed in the sweltering sun, surrounded by 'water, water everywhere, nor any drop to drink'. The mariner survives the ordeal, but the crew are not so lucky, falling prey to the ship of Death. In penance, the mariner is doomed to spend the rest of his life as a nomad, proselytizing on the dangers of environmental destruction.

The most enduring piece of imagery in the *Ancient Mariner* is that of the albatross, symbol of good fortune. But why was this bird thought to bring good luck? Basically, it was due to a misinterpretation. Sailors spent many weeks at sea, out of sight of land and dreaming of reaching port. Often one of the early signs that they would be making landfall in the near future was the sighting of birds, which indicated – like Noah's dove and its olive branch – that dry ground must be near by. The albatross, as one of the most noticeable birds on the planet (some have a wingspan of over 3.5 metres), was a major omen. The only problem is that the albatross, uniquely among birds, spends the majority of its life out at sea. Some birds have actually spent more than two years wandering around, often sleeping in flight as they glide effortlessly over thousands of kilometres of open ocean. So while the sailors thought they were seeing Noah's dove, they were in fact being duped by a peripatetic juggernaut.

The only problem with spending your life flying around the world's oceans is that, if you are a terrestrial species – even an amazingly adapted one like the albatross – you still need to return to land to have your babies. The albatross has a characteristically albatross-like solution to this problem, providing us with a fascinating bit of natural history. Despite its peripatetic lifestyle, and despite having a lifespan of over fifty years, the albatross always returns to the same island in order to mate. It mates for life, and its mate returns to the island as well, where they meet up to raise their single chick, splitting the chores equally. After a few months, when the young albatross is ready to head out into the world, they say their goodbyes, jot down the date of next year's rendezvous in their diaries and head back out to sea.

The evolutionary effect of always returning to the same island is that, while it encourages speciation between islands – with each island evolving into its own species over time – it tends to homogenize the birds that breed on any particular island. When the young albatrosses get together on their birth-island for the first time as adults, the males perform a ritual courtship dance to impress the females, who make their choice of mate without noting which part of the island the male hails from. As long as you are an albatross and you are on the island at the right time (natural selection takes a rather dim view of 'running a bit late' in this case), you've got a good chance of getting lucky.

The evolutionary term for a species like the albatross is panmictic – meaning that each individual has the potential to mate with any other individual in the species. While the albatross may fly over a significant part of the world's oceans during its lifetime, it doesn't put down roots anywhere but in its own home town. Humans aren't like this. When we move, we tend to mate with people living in the new neighbourhood. If we plot the distance between birthplaces of married couples over time, we see that until quite recently – the past hundred years or so – this distance was pretty small. My wife and I were born about as far apart as you can get – Atlanta, Georgia, and Hong Kong – but this would have been virtually unheard of a few generations ago. She would have ended up with someone living on Kowloon or the Mid-Levels, while I would have gotten hitched to a Southern belle.

The effect of this localization of mating habits is to make people living in the same region more similar to each other over time, and to increase the divergence between localities. If you met your third cousin, would you recognize him or her as a relative? If you didn't, and you hit it off and had a child together, what would that mean? Genetically, it would mean that your son or daughter would have slightly less than two unrelated parents, since you would share some of your genome with your mate. This means that the multiplier in our ancestor calculation would be less than two – providing us with the answer to our mathematical conundrum. Because people historically have tended to choose their mates from those living close by, they have inevitably ended up with someone they are related to – however distantly. This has the effect of making people living in the same region more similar to each other.

In some regions, of course, the degree of relatedness is quite high, with first-cousin marriages fairly common – we all have our favourite scapegoats for anecdotes about 'inbreeding'. But even if the degree of relatedness isn't high, over time the slight degree of inbreeding that has occurred in all traditional societies will tend to produce a distinctive pattern in the frequency of polymorphisms in that region. So, in the same way that you are uniquely defined by your polymorphisms as being the child of your parents, so too are people from a particular part of the world carrying a genetic signal of their geographic origin. It is these signals that we study as population geneticists – not simply the species unity of our common ancestors, Adam and Eve, shared by all of us, but the additional 'regional unities' that make up the patchwork quilt that is modern humanity. As we saw from Dick Lewontin's analysis, these signals are quite weak – but they are there. The trick is to find the polymorphisms that do unite us into regional groups, and to do this we need to spend a bit more time in the lab.

. . . nor any drop to drink

Zuckerkandl and Pauling's insight into diverging molecules as the timekeepers of evolution, and their utility for peering back into the past to see the common ancestor, gave us a clue about how to interpret the mass of mitochondrial data and infer the existence of Eve. Of course, since the Y-chromosome is also free from recombination, the same applies to it. By following the pathway defined by Y polymorphisms, we can reach Adam easily and quickly as well – all we need are the polymorphisms. And here the Y plays a trump card, because until quite recently it looked like there just weren't that many.

In 1994 Rob Dorit, Hiroshi Akashi and Walter Gilbert (the same person who co-discovered DNA sequencing in the 1970s) published an odd paper in the prestigious scientific journal *Science*. It was odd not because of what they had found, but because of what they hadn't. Titled 'Absence of polymorphism at the ZFY locus on the human Y-chromosome', it described an analysis of thirty-eight men from around the world as part of a focused effort to discover polymorphisms on their Y-chromosomes. Although a few polymorphisms had been identified on the Y – the first were discovered independently by Myriam Casanova and Gerard Lucotte in 1985 – there were far fewer than were known for any other chromosome. The surprising result of the Dorit survey was that there was no variation on the human Y-chromosome in the region examined. There was not a single DNA sequence variant detected, which implied that all of the men shared a very recent common ancestor. But since there was no variation detected, it was impossible to say when this person may have lived. On the face of it, they all could have had the same father – a Casanova of a man who had sown his oats all over the world. However, owing to the relatively small amount of DNA they studied – around 700 nucleotides in length – and the small number of men, it was also possible that they had simply been unlucky and chosen a region that didn't vary in those particular Y-chromosomes. For this reason, the estimate of the date of the most recent common ancestor of the men – in other words, Adam – was between 0 and 800,000 years ago. This provided no new insights into human origins and migrations, other than to serve as a deterrent to researchers who wanted to study the population genetics of the Y.

A few polymorphisms did turn up over the next few years, and Michael Hammer of the University of Arizona was able to find enough diversity to place Adam in Africa within the past 200,000 years – confirming the mitochondrial results and, tantalizingly, setting the stage for an ancestral tryst on the veldt. The total number of informative Y polymorphisms was still quite small, however. The time had come for a scaling-up of the search for diversity, and again, the San Francisco Bay area of California was to provide the right setting.

Under pressure

Peter Underhill started his scientific career studying marine biology in California in the late 1960s, ultimately obtaining a PhD from the University of Delaware in 1981. He then returned to California, taking a leap into the emerging field of biotechnology, doing things like designing enzymes for use in molecular biology research. Most importantly, he was absorbing the dizzying array of emerging technologies that geneticists were developing at the time. This was a heady time for the fledgling biotech industry, and the San Francisco area was the epicentre of the revolution promised by recombinant DNA. Cutting and splicing genes became the biological counterpart to the expanding computer industry in Silicon Valley and the surrounding towns.

In 1991, tired of the commercial world, he applied for a position as a research associate in Luca Cavalli-Sforza's laboratory at Stanford University. After convincing Luca that he would fit into the close-knit and collaborative group, he was hired. Peter started off in the lab by sequencing mtDNA, but he soon became interested in the Y-chromosome. The Cavalli-Sforza laboratory at that time was a very exciting place to be, with a real sense of 'blazing a new path' in the field – I count myself lucky to have been a postdoctoral fellow there at the time. New methods of statistical and genetic analysis were being developed almost weekly, and the intellectual climate was impeccable. Nearly all of the major figures in human population genetics spent some time at Stanford during the 1990s – among them students and postdoctoral research fellows such as David Goldstein, Mark Seielstad and Li Jin, all of whom we will encounter later in the book. But it was an analytical chemist, oddly enough, who was to have the greatest impact on our story. To explain why, we need to know a little bit about the molecule that makes up our genome.

One of the main tools in the geneticist's technical arsenal is the ability to separate fragments of DNA on the basis of size. The DNA inside your cells, like the proteins, is a linear chain of building blocks known as nucleotide bases. The information is encoded in the sequence of bases that make up DNA, rather like the amino acids that make up a protein. Unlike proteins, however, DNA has only four building blocks, called *nucleotide bases*: adenine (A), cytosine (C), guanine (G) and thymine (T). The information they encode – the instruction manual to build you – is contained in the particular sequence of these four nucleotides. In the same way that Morse code can convey a huge amount of information with only dots and dashes, so too can DNA encode the biological essence of an organism in the pattern of nucleotides. With 3 billion of them to work with, that's a lot of data.

Techniques that separate a mixture of molecules on the basis of their *size* can actually be used as a method of inferring the *sequence* of nucleotides in a DNA molecule. This is because biochemical techniques can generate DNA fragments of a particular length based on their sequence. After the fragments are generated, they can be separated by passing them through a gelatine-like matrix in the presence of an electric field. Because DNA is negatively charged, the fragments migrate toward the positively charged end of the matrix – at the molecular level, opposites really do attract. Interestingly, by doing this in a gel matrix the fragments will be retarded in their movement, because they have to navigate through the maze of tiny channels in the gel. The extent to which they are retarded depends on their length – long molecules are retarded to a greater degree than short ones, since they have more material to squeeze through the matrix channels. All very complicated in theory, but it works beautifully in practice. This technique, known as sequencing, is the basis of almost every important genetic discovery that has been made in the past thirty years. The sequencing of the human genome, for instance, involved the application of this technique tens of millions of times – not a terribly exciting task, but effective.

One problem with sequencing is that it is quite slow, and the biochemical reactions that allow you to determine the sequence of the DNA molecule you are studying can be very expensive. For this reason, geneticists try to use quicker and cheaper methods to examine DNA sequences, often looking for differences between a tested individual and one whose sequence has already been determined laboriously by the biochemistry and gel methods. The differences between the DNA sequences are our polymorphisms, and they help to determine individual susceptibility to disease, hair colour (assuming you haven't modified it) and all of the other inherited differences between people.

But most of them have no effect on the person carrying them – they are inherited baggage, markers of your ancestry. These are the markers of greatest interest to anthropologists and historians.

Peter Oefner, our chemist, is a serious, driven Austrian from the Tyrol region near Innsbruck. In the 1990s he was conducting research at Stanford on the separation of DNA molecules using a technique known as High Pressure Liquid Chromatography (HPLC for short). In particular, he was trying to develop a method of identifying the sequence of a DNA molecule using HPLC, which separates molecules much more quickly than gel methods. Peter Underhill saw Oefner's presentation on the technique at a noontime seminar in the Genetics department. Underhill was immediately struck by its applicability to the problem of finding Y-chromosome polymorphisms, and approached Oefner to ask if he would be interested in collaborating. The pair were soon in a frenzy of work that would see both of them give up their weekends for the next eighteen months.

The partnership between the two Peters would eventually produce a technique known as denaturing HPLC, or dHPLC for short. It makes use of a fortuitous property of DNA molecules: they are double-stranded, paired nucleotide chains held together by a mutual attraction between their constituent nucleotide bases. In the world of DNA, adenine always pairs with thymine, and cytosine always pairs with guanine, owing to the nature of their molecular structure. This means that if you know the sequence of nucleotides in one strand, then you automatically know that of the other strand as well. This has two knock-on effects. First, it stabilizes the DNA molecule, rendering it less susceptible to destruction by enzymes and environmental stress. DNA has been recovered from 50,000-year-old bones, but the single-stranded equivalent also found in our cells, known as RNA, is simply too unstable to last that long. The second benefit of being double-stranded is that it provides a way of backing up the data contained in the nucleotide sequence. If a change (i.e. a mutation) does occur on one strand of the DNA molecule, the mirror-image nucleotide on the opposite strand will no longer pair with it perfectly. There will be a slight 'kink' in the strand at this point, due to the mismatched base pairs. The kinks are easily detected by proofreading machinery in the cell, and the damage is repaired.

The technique of dHPLC uses the incredibly sensitive separation technique of HPLC as a substitute for the cellular proofreading machinery. It does this by passing the mismatched DNA molecules through a matrix that retards their movement based on the structure (but not the length) of the molecule. If there is a kink in the strand, the movement is altered, and the mismatched fragments can be detected by a different pattern of migration. This allows you to scan an entire DNA fragment – hundreds of nucleotides in length – for any differences between it and another DNA fragment of known sequence, quickly and cheaply. A fantastic time-saver and a critical leap forward in our ability to 'sequence' our genes.

The medical applications of this fancy bit of physical chemistry seem obvious, and the technique has been applied to determine the genetic mutations at the root of several human diseases. But what does it add to the study of ancient migrations? The answer is that, by applying this technique to the same region of DNA in many individuals, we can detect the genetic differences between them. This allows us to assay the level of genetic diversity in the human species rapidly and efficiently, providing a variety of polymorphisms to study. Before this technique was developed, there were perhaps a dozen polymorphisms identified on the Y-chromosome. At last count there were around 400, and the number is increasing weekly. If Rob Dorit and his colleagues had been able to perform their study of Y diversity with dHPLC, they would have found some variation. As often happens in science, technology has opened up a field to new ways of solving old riddles – often providing startling answers.

Adam's late

The obvious first question to ask is, do the large number of Y polymorphisms still indicate an African origin for modern humans? The unequivocal answer is yes, and a study published by the Peters and nineteen other authors (including myself) in the scientific journal *Nature Genetics* in November 2000 stated the results clearly and succinctly. A worldwide sample of men, from dozens of populations on every continent, were studied using the newly discovered treasure trove of Y polymorphisms. Applying the same methods used in the earlier mtDNA studies, a tree diagram was constructed from the pattern of sequence variation. What this diagram showed was that the oldest splits in the ancestry of the Y-chromosome occurred in Africa. In other words, the root of the male family tree was placed in Africa – exactly the same answer that mtDNA had given us for women. The shocker came when a date was estimated for the age of the oldest common ancestor. This man, from whom all men alive today ultimately derive their Y-chromosomes, lived 59,000 years ago. More than 80,000 years after that estimated for Eve! Did Adam and Eve never meet?

No they didn't, but the reason is fairly complicated, and it reveals one of the most important things to remember about the study of human history with genetic methods. When we sample people alive today, and examine their DNA to look for clues about their past, we are literally studying their genealogy – the history of their genes. As we have seen, people inherit their genes from their parents, so the study of genetic history is also a study of the history of the people carrying these genes. Ultimately, though, we hit a barrier when we trace back into the past beyond a few thousand generations – there is simply no more variation to tell us about these questions of very deep history. Once we reach this point, there is nothing more that human genetic variation can tell us about our ancestors. We all coalesce into a single genetic entity – 'Adam' in the case of the Y-chromosome, 'Eve' in the case of mtDNA – that existed for an unknowable period of time in the past. While this entity was a real person who lived at that time – the common ancestor of everyone alive today – we can't use genetic methods to say very much about *their* ancestors. We can ask questions about how Adam and Eve relate to other species (are humans more closely related, as a species, to chimpanzees or sturgeons?), but we cannot say anything about what happened to the human lineage itself prior to the coalescence point. Ockham's blade has nothing left to cut.

What this means for the estimate of coalescence dates is that, beyond placing all modern humans in Africa within the past 200,000 years, and therefore disproving the multiregional model of human evolution favoured by Coon and others, the dates have very little significance.

They do not represent the date of origin of our species – otherwise Eve would have been waiting a *long* time for Adam to show up. They simply represent the time, peering back into the past, when we stop seeing genetic diversity in our mtDNA and Y-chromosome lineages. Since mtDNA and the Y-chromosome are completely independent parts of our genetic tapestry, it is perhaps not terribly surprising that they coalesce at different times. Were your parents born on the same date, for instance? Also, the estimates of genetic dates – as with those of archaeological dates – involve some assumptions about past populations that may not be completely accurate, and thus there is a range of dates that we get from our calculation of Adam's age, between 40,000 and 140,000 years, with 59,000 years being the most *likely*. As we'll see in Chapter 8, the age difference between Adam and Eve is larger than we would expect by chance, and is probably the result of thousands of years of sexual politics. It is not, though, indicative of any deep uncertainties about human evolution. Referring back to our sojourn in Provence, men simply lose their soup recipes more quickly than women.

So, the main point to be inferred from our estimates of the age of the coalescent points – Adam and Eve – is that there were no modern humans living outside Africa prior to the latest date we can estimate. Given that the Y date is later, this means that all modern humans were in Africa until at least 60,000 years ago. That is the real shocker: 60,000 years may not seem very recent, but remember that we're dealing with evolutionary time scales here. Apes first appeared in the fossil record around 23 million years ago – a huge expanse of time, and difficult to envision. But if we compress it down to a year, it helps to place the other dates in context. Imagine, then, that apes appear on New Year's Day. In that case, our first hominid ancestors to walk upright – the first ape-men, in effect – would appear around the end of October. *Homo erectus*, who left Africa around 2 million years ago, would appear at the beginning of December. Modern humans wouldn't show up until around 28 December, and they wouldn't leave Africa until New Year's Eve! In an evolutionary eye-blink, a mere blip in the history of life on our planet, humans have left Africa and colonized the world.

Given how recent this date is, can we still see any evidence of these early humans in the Africans living there today?

The importance of clicking

One of the most interesting things to come out of the Y-chromosome analysis is the pattern of diversity within Africa, seen in the distribution of deep genetic lineages within the continent. While all African populations contain deeper evolutionary lineages than those found outside the continent, some populations retain traces of very ancient lineages indeed. These groups are found today in Ethiopia, Sudan and parts of eastern and southern Africa, and the genetic signal they contain is very good evidence that they are the remnants of one of the oldest human populations. The signals have been lost in other groups, but today these eastern and southern African groups still show a direct link back to the coalescence point – Adam.

The populations involved encompass the African Rift Valley, extending into south-western Africa, where people known as the San – formerly called Bushmen – have a very strong signal of the diversity that characterized the earliest human populations. They also speak one of the strangest languages on the planet, notable for its use of clicks as integrated parts of words – like the clicking sound we might make when we guide a horse, or imitate a dripping tap. No other language in the world uses clicks in regular word construction, and this quirk has inspired linguists to study the San language family for nearly 200 years, since Europeans first colonized southern Africa. The languages of the family are incredibly complicated. English, for example, has thirty-one distinguishable sounds used in everyday speech (two-thirds of the world's languages have between twenty and forty), while the San !Xu language (the '!' in !Xu sounds a bit like a bottle opening) has 141. While it is uncertain exactly which forces govern the acquisition of linguistic diversity, this figure is certainly suggestive of an ancient pedigree – in exactly the same way that genetic diversity accumulates to a greater extent over longer time periods.

The pattern of deep genetic lineages within the San is also seen for mitochondrial DNA, and the convergence of these three independent lines of evidence – Y, mtDNA and linguistic – strongly suggests that the San represent a direct link back to our earliest human ancestors. Does this mean that our species originated in southern Africa, rather than the Rift Valley? Not necessarily, although the importance of our southern hominid ancestors has increased in recent years, and some palaeoanthropologists now argue for a southern genesis. What is clear is that the current distribution of the San people is a small portion of their historical range, and skeletal material classified as San-like has been unearthed from Palaeolithic sites in Somalia and Ethiopia. Some of the clearest modern evidence for this again comes from linguistics. Outside southern Africa, the only other place where click languages are spoken is in east Africa. The Hadza and Sandawe of Tanzania speak very divergent click languages, providing evidence for a once widespread linguistic family stretching from the Rift Valley to Namibia. It is likely that this continuous distribution was overrun relatively recently by the migrations of Bantu-speaking populations from central Africa, who expanded over much of eastern and southern Africa in the past 2,000 years. Prior to the coming of the Bantus, however, southern and eastern Africa appears to have been predominantly San.

Face to face

One of the distinguishing features of the San people is their 'non-African' physical appearance. Of course, there is tremendous diversity of physical appearance in Africa, and any attempt to classify people according to African and non-African type is meaningless. When most of us think of Africans, we tend to picture the typically Bantu features of central Africans and (via the European slave trade) of African-Americans and Afro-Caribbeans. The San are a much smaller people, with lighter skin, more tightly curled hair and a thicker layer of skin over the eyes – the so-called epicanthic fold that also characterizes people from east Asia. It is this latter feature which has led some researchers to suggest that the epicanthic fold is an ancestral characteristic of our species, and was simply lost in western Eurasian and Bantu populations. This hypothesis remains purely speculative until the genetic basis of the epicanthic fold has been deciphered, but it is certainly consistent with the evidence from the San. So do the San give us a glimpse of our ancestors who lived at the time of our genetic Adam?

It is difficult to imagine what our common male and female ancestors would have looked like. We can only make informed guesses, based on the diversity we see in human populations today, and informed by our perceptions of the processes of human morphological evolution. In this sense, it is like any historical science, where we base our understanding of an unknown past event on the extant clues – cutting through the complexity with the power of parsimony. Unfortunately, we have no real way to evaluate the accuracy of the likenesses produced, so some of this will have to be taken on faith.

It is unlikely that our African ancestors were the hairy, brutish troglodytes portrayed in museums – these are probably overly influenced by our perception of Neanderthals, who may have been pretty hairy and brutish. Rather, they are likely to have been fairly gracile and elegant, at least in comparison to Neanderthals. The simple reason is that the great mass of a Neanderthal, and the likely hairy exterior, is thought to have been an adaptation to the cold Eurasian climate. Because our earliest ancestors lived in the relatively warm climes of southern and eastern Africa, they would not have needed the warmth provided by a furry exterior.

They probably had the epicanthic fold. While this feature could have arisen twice in different parts of the world, it is more likely to have been a characteristic found in our common ancestors which was simply lost in the lineages leading to central and western Eurasians. Of course, the epicanthic fold arises *de novo* in every case of Down's syndrome, so perhaps it is relatively easy to create. A good working hypothesis, though, is that it is an ancestral feature.

Early humans probably had fairly dark skin. This is because of the nature of the environment where they lived – a sunny African savannah – where the protection against solar radiation afforded by dark skin would have been a distinct advantage. It is also because at least some of the mutations that produce light skin colour in Europeans and north-east Asians are derived from the ancestral, darker form of the gene (known as $MC1R$, or *melanocortin receptor*), which is virtually the only form found in Africa today. Thus, it seems likely that Africans have retained a darker colour, rather than evolving it from a lighter form.

Our ancestors of 60,000 years ago were probably about the same height as you and I – although this is really a meaningless statement. The average height of modern humans varies greatly around the world, with the Dutch being the tallest European population – young men are, on average, over six feet (1.83 metres) tall, and women are a few inches shorter. The Japanese are somewhat smaller, with men standing around 5 feet 6 inches high (1.7 metres). The Twa pygmies of central Africa, however, are significantly shorter – males are only 5 feet (1.5 metres) on average. This variation in stature probably reflects adaptations to local environments, which can be seen in our ancestors *Homo erectus* and *Homo ergaster* as well.

So, the picture that emerges is of a dark-skinned (although perhaps not as dark as some Africans today), reasonably tall, thin person – perhaps with an epicanthic fold. Someone who wouldn't look that out of place today dressed in a suit and sitting opposite you on the train. Not surprising, I suppose, given that he only lived about 2,500 generations ago.

Out of the nest

Accepting the evidence at face value, the implication is that Adam lived in population groups directly ancestral to the modern San, in eastern and/or southern Africa, around 60,000 years ago. The date of the earliest modern human populations – the first of our species – remains to be assessed, and could be anywhere between 60,000 and several hundred thousand years ago. We simply lose the signal from our genes at that stage, as all of the genetic diversity present today coalesces to a single ancestor. What is clearly implied by the data, however, is that all modern human genetic diversity found around the world was in Africa around 60,000 years ago. The mtDNA and Y-chromosome give us the same dates for the earliest non-African genetic lineages, and it is now agreed by most geneticists that humans began to leave Africa around this time. There may have been occasional forays into the Middle East prior to this, as suggested by 100,000-year-old human remains at sites such as Qafzeh and Skuhl in present-day Israel, but the Levant of 100,000–150,000 years ago was essentially an extension of north-eastern Africa, and was probably part of the original range of early *Homo sapiens*. The real expansion was beyond the Mediterranean world, into the uncharted territory of Asia proper.

Here we run headlong into what the Australians might call 'a curly one'. According to the dated remains in Australia, humans were there, 15,000 km east of Africa by the shortest land route, at the same time we are all supposed to have been in Africa, 50–60,000 years ago. If I were prone to bouts of mysticism, I might infer from this that the ancestors of the Aborigines had learned how to 'fold space', as Frank Herbert called it in the science fiction novel *Dune*. Being (reasonably) firmly grounded in the pragmatic and rational world of science, however, I am forced to look elsewhere for answers.

4
Coasting Away

So it was, on this First Morning, that each drowsing Ancestor felt the Sun's warmth pressing on his eyelids, and felt his body giving birth to children. The Snake Man felt snakes slithering out of his navel, the Cockatoo Man felt feathers. The Witchety Grub Man felt a wriggling, the Honey-ant a tickling, the Honeysuckle felt his leaves and flowers unfurling. The Bandicoot Man felt baby bandicoots seething from under his armpits. Every one of the living things, each at is own separate birthplace, reaching up for the light of day.
Bruce Chatwin, *The Songlines*

When I was a child, my friends and I used to play a silly quiz game with each other, where we would ask trick questions intended to show off our command of obscure facts. One of the favourites was to name the largest island on earth. The naïve answer was 'Australia', which would always elicit a groan of disapproval. This is because Australia, as the groaners knew, is part of the continent of Australasia – not simply a large island. Encompassing Australia, New Zealand, Tasmania, New Guinea and several of the easternmost Indonesian islands, Australasia is the 'odd man out' in the geographic mnemonic stakes. And what an odd continent it is.

Present-day Australia is the driest subcontinent on earth – more than 90 per cent of it receives less than 1,000 mm of rainfall per year. Partly as a response to the environmental challenge of living there, it is the most urbanized nation in the modern world, with 90 per cent of its population living in cities along the coast. It boasts the planet's longest continuous coral reef, the awe-inspiring 2,000-kilometre Great Barrier Reef. Perhaps the most interesting thing about Australia,

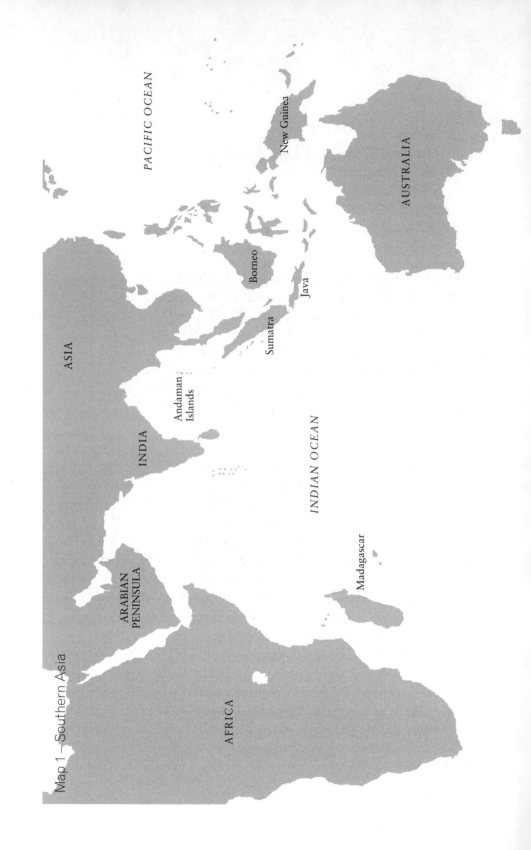

Map 1 — Southern Asia

PACIFIC OCEAN

New Guinea

AUSTRALIA

ASIA

Borneo

Java

Sumatra

INDIA

Andaman
Islands

INDIAN OCEAN

ARABIAN
PENINSULA

Madagascar

AFRICA

though, is its fauna. The animals in Australia are unlike those anywhere else on the planet, with the only similarities found to those of New Guinea – also part of Australasia. The reason for this uniqueness is the extreme isolation of the place. Anyone who has sat through the two-night flight from London to Sydney can attest to the difficulty involved in getting there. Through the vagaries of plate tectonics, Australia has been disconnected from the continents of Eurasia, the Americas and Africa for the past 100 million years or so – its most recent connection was to Antarctica! What this isolation has meant is that Australia has missed out on most of the main line of mammalian evolution, with its wealth of placental species. The lack of 'normal' mammals has allowed evolution to pursue a different path, resulting in oddities like the platypus and the kangaroo. It has also meant that, until quite recently, Australasia had no primates – no monkeys, no apes, not even a bushbaby. Humans are the only primate species on the continent.

The lack of evolutionary antecedents means that humans must have colonized Australia from somewhere else. But where did they come from? The journey clearly involved a significant sea voyage, even from its nearest continental neighbours. If we allow for fluctuations in sea level as a result of climatic fluctuation, the landmass of Sahul (which included New Guinea and Tasmania, in addition to Australia) that was created during the coldest part of the last ice age, approximately 20,000 years ago, would still have been approximately 100 km from the rest of south-east Asia. The answer to how and when Australia was colonized by humans is one of the key pieces in our effort to solve the puzzle of how modern humans settled the world. The details it reveals about human history – and the methods of analysis involved in piecing it together – will set the pattern for the rest of our journey.

Death and decay

Lake Mungo, in New South Wales, is about 1,000 km west of Sydney. From the nearest town with an airport, Mildura, it is a 120-km drive on a dirt track through the hot scrub desert that comprises much of inland Australia. Mungo is no longer a lake – the water dried up over

10,000 years ago, leaving behind fantastic sand and clay formations that are reminiscent of those at Mono Lake in northern California – but between 45,000 and 20,000 years ago it was part of a lush oasis known as the Willandra Lakes. The lakes were fed by the Willandra Creek, which joined the Murray River further south, and ultimately emptied into Encounter Bay near present-day Adelaide. From the animal remains found at the site it is clear that several large species of extinct marsupials lived around the lakes, including the buffalo-sized *Zygomaturus* and a 200-kg short-faced kangaroo, *Procoptodon*. All of the animals of the area were herbivores, and as such they would have been tempting prey to humans.

It was around the earlier end of this time range, according to recently obtained dates, that a man was buried there. Called Mungo 3 by his discoverer, Jim Bowler, the find was dated to around 30,000 years ago when it was discovered in 1974. More recent dating methods have pushed the age back to 45,000 years, and human artefacts from sedimentary layers below Mungo 3 hint at dates as ancient as 60,000 years before present. If confirmed, these dates will make Mungo the earliest site in the world outside Africa to be inhabited by anatomically modern humans.

The earliest human remains in Australia, like those elsewhere in the world, have been dated using isotopic decay methods. These methods measure the ratio of different isotopes of an atom present in the sample. It is possible to do this because almost all atoms come in more than one 'flavour', depending on how many subatomic building blocks (particles called neutrons) they have. Through the alchemy of particle physics, the 'heavier' atoms tend to shed some of their particles over time, in the process transforming them into the 'lighter' atoms. By knowing the rate at which this decay occurs, and measuring the ratio of the heavier to the lighter atom, it is possible to calculate how long the decay has been going on. Like the molecular clock discussed in Chapter 2, this atomic clock provides critical time estimates for the study of ancient human remains.

The most widely applied form of isotopic dating is the so-called radiocarbon method, which measures the ratio of Carbon-14 (C-14) to Carbon-12 (C-12) in the sample. C-14, through a complex inter-action with the atmosphere, breaks down to Nitrogen-14 (N-14). The rate of breakdown depends on the so-called half-life of C-14, which is the amount of time required for one-half of the C-14 in a sample to decay – around 5,700 years. Since carbon is used to build organic molecules, like those found in plant and animal tissues, the method is fantastic for dating human remains. The problem is that beyond about 40,000 years ago, the estimates of C-14:C-12 ratios are not terribly accurate, since most of the C-14 has already decayed. After 5,700 years, only half of the C-14 originally incorporated into the tissue when the organism was alive is still there, and after 11,400 years only a quarter is still present. By the time we get to 40,000 years, only one sixty-fourth of the original C-14 is still present – less than 2 per cent. This makes the sample extremely susceptible to contamination by minute quantities of modern material, which would have the effect of making the dates appear to be more recent than they actually are. For this reason, radiocarbon dating tends to be most useful for remains that are younger than around 30,000 years, and it is the method of choice for archaeological sites of the past 10,000 years, where it is extremely accurate.

Once we get beyond 40,000 years, though, we have to use isotopes that decay at a slower rate. Two of these are Potassium-40 (K-40) and Uranium-238, which have half-lives of 1.25 billion and 4 billion years respectively. The problem with the more stable isotopes is that they are not usually found in the stones and bones themselves, and therefore they can be applied only to the sediments surrounding the remains – typically volcanic ash in the case of the former and lake sediments in the case of the latter. Thus, you have to have been very lucky with your choice of sites to be able to use them. Thankfully, the geological activity of Africa's Rift Valley has meant that K-40 dating can be widely applied there.

But what if you aren't so lucky? In particular, what if your remains are beyond the useful limit of radiocarbon dating, but they aren't found in sediments that allow you to use the other methods? Then we have to rely on a collection of three relatively new techniques in the isotopic arsenal called – rather intimidatingly – thermoluminescence, optically stimulated luminescence and electron spin resonance. All rely on the observation that naturally occurring radiation causes electrons – another type of subatomic particle – to accumulate in small crystalline defects in a substance at a steady rate, depending on the level of exposure to an 'electron-cleansing' radiation source such as fire or sunlight. There are many assumptions about the degree to which electrons had accumulated in the defects, known as traps, before being exposed to the cleansing radiation source. Also, there are assumptions about the variability in radiation exposure over time. For these reasons the dates obtained using luminescence and resonance methods are not as accurate as those obtained with C-14 or K-40 dating. However, for many sites they are the only option available.

It is exactly these last techniques which have been most widely applied in Australia. In particular, several objects obviously manufactured by humans – some of them associated with artistic depictions on rock faces – have been dated to more than 40,000 years ago. Of course, with the uncertainty of the techniques, it is difficult to know how accurate these dates are. But there is evidence from other sources that humans have been in Australia for a very long time indeed. Richard Roberts and his colleagues at the Australian National University, investigating the relatively unsophisticated tools used by these early people, have inferred dates as great as 60,000 years ago for one site in the Northern Territory.

The weight of palaeoanthropological evidence is now clearly in favour of a very early settlement of Australia by modern humans – perhaps as early as 60,000 years ago. But the earliest archaeological sites on the south-east Asian mainland date to less than 40,000 years ago. How could humans have been in Australia 20,000 years before this – surely they came from south-east Asia? The answer to this conundrum will take us back to Africa, where we need to pay a visit to the Garden of Eden.

Surf and turf

Africa is the most equatorial continent on earth. The entirety of its landmass is found between latitudes of 38°N and 34°S, and 85 per cent of its land area is in the tropical zone between Cancer and Capricorn. Sea-level freezing temperatures are rare in Africa – uniquely among all the continents. While the interior deserts of the Sahara and the high volcanic mountains of east Africa are inhospitable to humans, most of the continent is surprisingly benign. Africa contains the Old World's largest uninterrupted tract of rainforest, and the savannahs of the east and south support a huge variety of large mammals. The combination of rainforest and savannah in close proximity, again unique in the Old World, is probably part of the reason that humans evolved there. Hominid bipedalism was almost certainly an early adaptation to the treeless grasslands of Africa, perhaps as early as 5 million years ago, where more resources could be exploited by leaving the aerial safety of the deep forests.

Africa was not always in the location it occupies today. Through the vagaries of plate tectonics, it spent most of its time between 200 and 20 million years ago migrating around the southern Indian Ocean, eventually bumping into the Eurasian landmass around 15 million years ago. It was at this time that the great apes began to disperse around the world as part of the first 'African Exodus'. Those that went east evolved into the orang-utan and gibbon – the species favoured by Eugène Dubois as our most likely ancestors. The apes that stayed evolved into the chimpanzee and gorilla – and eventually, perhaps 100,000 to 200,000 years ago, into anatomically modern humans. During this entire sequence, Africa remained in the same position geographically. But, as with the other continents, the climate has fluctuated dramatically in the past few hundred thousand years.

The field of palaeoclimatology investigates the climate of bygone eras. The earth at 150,000 years ago was nearing the end of what is known as the Riss glaciation. On average, the temperatures were 10°C colder than they are today, although there was substantial variation among the continents. Around 130,000 years ago it started to warm up, and tropical Africa began to get more rain as the sea levels rose and moisture re-entered the atmosphere. A period of gradual cooling began around 120,000 years ago, accelerating after 70,000 years ago. This pattern would continue (with short-term fluctuations) for the next 50,000 years, reaching its nadir around 20,000 years ago.

Because Africa is largely tropical, its climate depends less on the variation in solar intensity that produces seasons in higher latitudes. African weather patterns are largely determined by rainfall, with pronounced wet and dry seasons setting the tempo of life throughout the continent. The famous migration of the wildebeest in Kenya and Tanzania, for example, is triggered by the onset of the dry season in June. But the seasons have not always occurred with the same intensity, producing a climate that, in the past, has sometimes been wetter and sometimes drier than that today. These long-term fluctuations have almost certainly affected the movements of animals – including humans.

Recent research by Robert Walter, an American geophysicist based in Mexico, suggests that a large-scale drying up of the African continent at the onset of the last ice age resulted in modern humans favouring coastal environments. This is because savannahs are unusual places. They are closely related to tropical forests in the chain of climatic relationships, and the two zones are interchangeable depending on the level of rainfall. In general, the areas of tropical Africa with more than three months of low rainfall are savannah, while those with fewer than three are forest. If there are substantially longer dry periods, the environment grades into steppe, and ultimately into desert as moisture becomes extremely scarce. While these regions are all found in particular locations in present-day Africa, their past extent has fluctuated. What Walter's research suggests is that as Africa began to dry up, the savannahs of eastern Africa were replaced by steppe and desert, except in a narrow zone near the coast. It was in these coastal savannah environments that early humans would have congregated, exploiting food sources from the sea as well as those of the land animals living near by.

While the universality of this theory is uncertain, and it may turn out to be a minor sideline of human evolution, one thing is clear: there is incontrovertible proof that early humans were able to live off of the sea. Large middens, or garbage dumps, of shells from clams and oysters have been found in Eritrea, on the eastern Horn of Africa, dating from around 125,000 years ago. These middens also have human stone tools interspersed among them, showing that humans were living in the region and exploiting coastal resources. The presence of butchered remains of rhinoceros, elephant and other large mammals conjures up a prehistoric 'surf 'n' turf' feast reminiscent of the massive platters of steak and shellfish served in American restaurants. It seems that our distant ancestors had quite well-developed palates, even in those days of apparent hardship.

One of the most exciting details to emerge from Walter's work is the fact that there appears to have been exchange with coastal dwellers thousands of kilometres away, who were exploiting the same types of resources in southern Africa. This is suggested by the similarities in tools found at the sites, coupled with their roughly contemporary dates. It seems that humans were able to migrate over long distances, relatively rapidly, by following the coast of eastern Africa.

Now for the big leap: if humans could migrate over long distances within a continent, using the same technologies and exploiting the same resources, why couldn't they do the same between continents? The coastal route would be a sort of prehistoric superhighway, allowing a high degree of mobility without requiring the complex adaptations to new environments that would be necessary on an inland route. The resources exploited in Eritrea would be pretty much the same as those in coastal Arabia, or western India, or south-east Asia, or – wait for it – Australia. And because of the ease of movement afforded by the coast, the line of sandy highway circumnavigating the continents, this would allow relatively rapid migration. No mountain ranges or great deserts to cross, no need to develop new toolkits or protective clothing, and no drastic fluctuations in food availability. Overall, the coastal route seems infinitely preferable to anything further inland. There were only a couple of sections of open water that would have required a boat to cross. Most likely these boats would have been rather simple – probably a few logs lashed together – but we have no direct evidence, because wood disintegrates very quickly. Nevertheless, they did make it across.

It is clearly plausible that the early presence of humans in Australia, almost immediately after they left Africa, can be accounted for by migration along this coastal route – beachcombing along the southern coast of Asia. There are two remaining pieces of the puzzle to be evaluated, though – rather critical ones, in fact. If one of the early waves of migration out of Africa followed a coastal route, is there a telltale genetic pattern? It depends on the way in which the migration occurred, and what the migrants did along the way. We might expect to see a band of particular genetic markers along the coast, differentiated from the populations living further inland. Or perhaps the signals have been homogenized among descendants of the coastal dwellers and the land migrants. The only way to find out is to examine populations from along the route and see what the genetic pattern is. The second critical piece of evidence is to be found in the pattern of archaeological remains along the route – are they consistent with such a journey?

M&Ms

Mitochondrial DNA and the Y-chromosome, as we saw before, display deeper lineages within Africa than outside. What does this really mean? If we imagine the genetic relationships among modern mitochondrial diversity as an actual tree – say a large oak – then the root and trunk, and the branches that are closest to the ground, are all found in Africans. These branches sprouted first, as the tree was growing, and they are therefore the oldest. This means that the tree started growing in Africa. As we move further up the trunk, branches start to appear that are found in non-Africans. These formed later. How far up do we have to go before we find the non-Africans? The answer is pretty high. If the tree started growing 150,000 years ago – the age of the root – then the non-African branches are much closer to the top, and do not pre-date 60,000 years. Most of human evolution has been spent in Africa, so it makes sense that there is greater diversity there. Most of the branches on the tree are found only in Africans.

The beauty of the genetic data is that it gives us a clear, stepwise progression out of Africa into Eurasia and the Americas. The diversity we find around the world is divided into discrete, although related, units, defined by markers – the descendants of ancient mutation events. By mapping these markers on to the map of the world, we can infer details of past migrations. Following the order in which the mutations occurred, and estimating the date and any demographic details (such as population crashes or expansions), we can gain an insight into the details of the journey. And the first piece of evidence comes from one man in particular, who had a rather important, random mutation on his Y-chromosome between 31,000 and 79,000 years ago. He has

been named, rather prosaically, M168. More evocatively, he could be
seen as the Eurasian Adam – the great . . . great-grandfather of every
non-African man alive today. The journeys taken by his sons and
grandsons defined the subsequent course of human history.

It is perhaps surprising that the clearest evidence for the route
followed by our ancestors on their journey out of Africa comes from
the Y-chromosome – surely men tend to 'sow their oats', causing the
widespread dispersal of regional genetic signals? Oddly enough, no –
and the rapid loss of ancient soup recipes on the male lineage (which
we used to explain Adam's recent date) means that men living in a
particular area tend to share a recent common ancestor, providing us
with clear 'fingerprints' of particular geographic regions. What this
means is that the Y gives us the clearest evidence for the journeys
followed by early humans. It is literally a 'journey of man', but it is
the best tool we have for inferring the details of the trip. It is obviously
important to examine the female lineage to see if it follows the same
pattern – to make sure the fish stays with the bicycle, so to speak – but
the Y-chromosome does provide us with the cleanest distillation of
human migrational history.

As we look more carefully at the arrangement of branches on the
mitochondrial tree, we find that there is a similar pattern – all of the
non-African mitochondrial branches descend from a particular branch
of the tree trunk, implying that our M168 Adam was paired with an
Eve. Thankfully, this Eurasian Eve lived around 50–60,000 years ago,
suggesting that she and Eurasian Adam could have met. She is called
by the (again) rather mundane name L3, and her daughters accompanied the sons of M168 on their journey to populate the world.

Based on the distribution of the descendants of M168 and L3 in
Africa today, it is likely that they both lived in north-east Africa, in
the region of present-day Ethiopia and Sudan. Like all men alive today,
M168 shared deeper roots with his African cousins. His lineage is a
major branch leading off the human family tree, with his descendant
'terminal branches' found in the DNA of all of today's Eurasians, but
he connects them back through M168 to our species' African root. In
our tree metaphor, each marker that we study defines a node on the
tree – a point where a branch splits into two smaller branches. If we had
no markers apart from M168 and L3, our trees would be fairly sparse,

comprising a root (Adam and Eve) and one split on the tree, defined by M168 or L3, on the branch leading out of Africa, and another branch remaining in Africa. Luckily, the tree is packed with dense foliage, defining a pattern of growth that traces the map of our journey.

Intriguingly, on both the mitochondrial and Y branches, there is another split, immediately after M168 and L3, dividing the Eurasian branch structure into distinct clusters – two in the case of mtDNA, and three in the case of the Y*. For both the Y-chromosome and mtDNA, one cluster is more common than the other(s), accounting for around 60 per cent of the non-African branches (or lineages) in the case of mitochondrial DNA, and more than 90 per cent in the case of the Y. In other words, the majority of non-Africans alive today have mtDNA and Y-chromosomes belonging to the more numerous clusters – people living all over the world, in places as disparate as Europe, India and South America. The rarer lineages, though, are found only in Asia, Australasia and the Americas. It is these rare lineages that constitute the majority of the mitochondrial and Y types in the Australian Aborigines.

Our rare mitochondrial cluster is given the name M – like the head of M16 in James Bond movies. In biblical terms, Eve begat L3, and L3 begat M. According to recent research by Lluis Quintana-Murci, a Catalan researcher working in Paris, the distribution of the M cluster is indicative of an early migration out of Africa, which proceeded along the coast of south Asia, ultimately reaching south-east Asia and Australia. M is virtually absent from the Middle East, and is not found at all in Europe, but it constitutes 20 per cent or more of the mitochondrial types in India, and close to 100 per cent of those in Australia. Quintana-Murci estimates its age to be 50–60,000 years, and from its distribution it seems that people who carried the M lineage never made it into the interior parts of the Middle East. The most

---

* Of the three Y clusters tracing their descent from M168, only two will be examined in this book. The third, found mainly in Africa, is defined by a marker known as YAP or M1. Outside Africa it splits into two lineages, tracing essentially the same migrational routes as the other two Y clusters described in the text. Because it adds little to our understanding of the 'out of Africa' migrations, and is rare in most non-African populations, I have chosen to ignore it. My apologies to Mike Hammer, who discovered the YAP marker in the early 1990s.

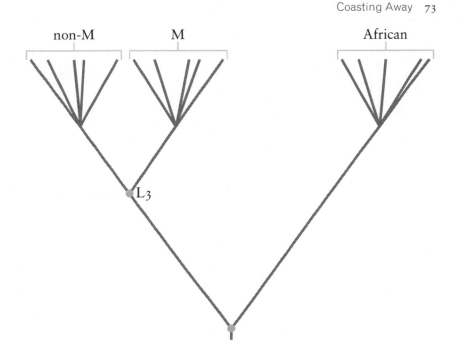

Figure 3 MtDNA genealogy, showing the split into M and non-M lineages outside Africa.

likely explanation is that the 'M people' left Africa very early on, carrying their distinctive genetic signature across the south of the continent along the coastal highway.

And what about the Y? Is there a male counterpart to our M mitochondrial lineage? Luckily, the answer is yes. Again assuming biblical style, Adam begat M168, and M168 begat M130. M130 appears to have accompanied mitochondrial M on her coastal journey, and the present-day distribution of his descendants provides us with an insight into the nature of the trip. Like the M mitochondrial lineages, M130 Y-chromosomes are limited to Asia and America, but the dynamics of lineage extinction that we see for the Y have left a much more striking pattern than the one seen for their mitochondrial counterparts. M130's descendants are virtually unknown west of the Caspian Sea, but they comprise a substantial proportion of the men living in Australasia. M130 is only found at low frequency in the Indian subcontinent – 5 per cent or less. But as we move further east,

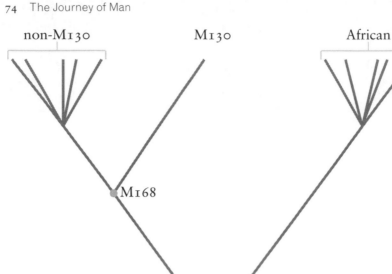

Figure 4 Y-chromosome genealogy, showing the split into M130 and non-M130 lineages from an M168 ancestor.

the frequency increases: 10 per cent of Malaysian, 15 per cent of New Guinean and 60 per cent of Australian aboriginal men trace their ancestry directly to M130. There is a quirkily high frequency of M130 in north-east Asia, particularly in Mongolia and eastern Siberia, which suggests a later migration that we will revisit in Chapter 7. For the purposes of our Australian story, though, M130 provides us with a clear fingerprint of the coastal migration out of Africa.

One other piece of evidence suggests a direct link between Africa and Australasia – physical appearance. The dark skin of the Australians is reminiscent of that found in Africa – something that begs an explanation. Most of the people living in south-east Asia today would be classified as 'Mongoloid' peoples, implying a shared history with those living further north in China and Siberia. There are, however, isolated populations of so-called Negritos living throughout south-east Asia who closely resemble Africans. The most obvious examples are found in the Andaman Islands, under the jurisdiction of India but actually 400 km off the west coast of Thailand. The largest tribal groups, known as the Onge and Jarawa, have many features that link them with the Bushmen and Pygmies of Africa, including short stature, dark skin, tightly curled hair and epicanthic folds. Other Negrito groups, such as the Semang of Malaysia and the Aeta of the Philippines, have mixed substantially with Mongoloid groups and have a more 'Asian' appearance. The Andamanese, probably because of their island home, have escaped much of the admixture seen on the mainland. Because of this they are thought to represent a relic of the pre-Mongoloid population of south-east Asia – 'living fossils', if you will. The suggestion made by many anthropologists, particularly Peter Bellwood of the Australian National University, is that the population of south-east Asia prior to 6,000 years ago was composed largely of groups of hunter-gatherers very similar to modern Negritos. Migrations from north-east Asia over the past few millennia have erased the evidence of these early south-east Asians, except in the case of small groups living deep in the jungles or – in the case of the Andamanese – on remote islands.

So, both the Y-chromosome and the mtDNA paint a clear picture of a coastal leap from Africa to south-east Asia, and onward to Australia. Taking the genetic dates as a guide, modern humans could have made this journey around the same time as the earliest archaeological evidence pointing to human occupation in Australia. DNA has given us a glimpse of the voyage, which almost certainly followed a coastal route via India. But is there any archaeological trace of this journey along the route?

A swim in Ceylon

This brings us back to the issue of the dating, particularly as applied to the Australian remains. No evidence for other hominids has ever been found in Australia – *Homo erectus* did not make it across the long stretches of open ocean that separated it from south-east Asia, despite living only a few hundred kilometres away in Java. Because *Homo sapiens* is the only hominid species that has ever been found in Australia, any evidence of human occupation sticks out like a proverbial sore thumb. Stone tools unearthed in Arnhem Land could only have come from one source – us. And if the radiometric dates say that stone tools were present in Australia 50–60,000 years ago, almost immediately after the genetic dates show us that our ancestors were still in Africa, this means that modern humans must have made use of a route that afforded extremely rapid movement. The coastal super-highway seems to be the most likely one.

As we have seen, though, there were other hominids living along the route followed by these beach dwellers. They also made stone tools, and these have been found throughout Eurasia. The easternmost exten-sion of the range of *Homo erectus* was Java, and it is possible that they even survived until around 40–50,000 years ago – long enough for the coastal migrants to have encountered them as they moved through the Indonesian archipelago. It is clear, though, that they must have become extinct almost immediately after the arrival of the Moderns, if not before. What is uncertain is whether we actively forced them out of the picture – a genocidal scenario that we will explore in greater detail when we get to Europe later in the book.

In the same way that extinct hominid species can be recognized by the size and shape of their bones, so too can tools and other artefacts be classified according to their style. I like to draw a parallel with the evolution of that icon of American culture, the Coca-Cola bottle, during the twentieth century. During the first seventy years of the century, the bottles were 8-ounce glass sculptures, with a distinctive curved shape that still evokes 1950s soda fountains and drive-ins. During the 1970s, a larger, lightweight plastic bottle was introduced to supermarkets – but it retained its hourglass shape, as though harking back to the bygone era of the earlier version. By the 1980s, though, the curves had been dropped in favour of a standardized, flat-profile plastic bottle now used by all drinks manufacturers. There is minor size variation – 2-litre juggernauts are common in the UK and America, while continental Europe opts for a slightly more elegant 1.5 litres – but the new style has become universal.

This progression of a universal form is seen for all human tools, from hammers to knives to guns to sauté pans. Everything evolves over time, and the most efficient form finds the most widespread application. It quickly dominates over competing forms, and eventually makes it difficult to remember the styles that were in use before. Even before the current era of globalization, the world had its 'killer apps' that dominated everything else. In the case of the period we are talking about, 50–60,000 years ago, the killer apps are grouped into a common cultural phenomenon known as the Late Stone Age, or more technically, the Upper Palaeolithic. The tools of the Upper Palaeolithic mark a radical departure from those that pre-date them, and are clear evidence for the presence of anatomically modern humans, as opposed to *Homo erectus* or Neanderthals, who remained trapped in a Middle Palaeolithic time-warp.

The details of the Middle to Upper Palaeolithic transition will be examined in the next chapter, but for the purpose of the story of our Australian coastal dwellers it is sufficient to say that the earliest Upper Palaeolithic tools mark the initial migration of modern humans into any geographic region. And that is why India is unusual, since there is actually very little evidence of the Upper Palaeolithic there. There is a general dearth of human remains from all periods leading up to the Upper Palaeolithic, but at least there are abundant tools from the earlier periods. The Upper Palaeolithic provides no telltale signs until very late in the day, and even then they show up in an unexpected place.

Fa Hien cave in Sri Lanka provides us with the earliest sign of the Upper Palaeolithic in the Indian subcontinent. The date, however, is a problem – the earliest clearly modern artefacts date from no earlier than 31,000 years ago. Nearby Batadomba Lena cave contains the earliest skeletal material from anatomically modern humans, also dating from around 30,000 years ago. The combination of age and location gives us two clues in our search for traces of the coastal migration. First, the Sri Lankan caves suggest that the earliest modern humans arrived in India from the south, rather than from the north via the more obvious inland route. This implies that they were living on the coast, consistent with the theory of an early coastal migration.

The second clue, which comes from the date, is that the Batadomba people could not have been the ancestors of the Australians, since they actually lived over 20,000 years after the earliest evidence for human settlement in Arnhem Land. Another curly one. It may turn out that archaeological layers below those already excavated will yield earlier evidence for modern human presence, but for the time being it appears that Batadomba is too late to help us along on our voyage. In fact, late dates are found along the entirety of our coastal route to Oz. In Thailand, for instance, there is evidence for modern human occupation from about 37,000 years ago at Lang Rongrien cave – but not before. As we move closer to the scene of the crime, the dates get older – advanced, Upper Palaeolithic stone tools dating from 40,000 years ago have been found at Bobongara, on the Huon Peninsula of eastern New Guinea. This would have been the final stepping-stone on the journey, but there is still nothing approaching the 50–60,000-year-old dates in Australia. Thus, in spite of the genetic pattern tracing an early coastal route out of Africa, the archaeology appears to have failed us. Where is the evidence for our coastal route?

Unfortunately we don't know, but there is a likely hypothesis. Since almost all archaeological work today is carried out on land, we are probably missing the artefacts that are hidden underwater. 'Rubbish – surely Atlantis is a myth!' you might be saying. Well, yes and no. While the evidence for an entire civilization falling catastrophically into the sea is fairly sparse, what is unequivocal is that sea levels have indeed fluctuated substantially – if somewhat more gradually – over the past 100,000 years. Those of 50,000 years ago were around 100 metres lower than they are today, as large amounts of moisture were tied up in the expanding ice sheets of the northern hemisphere. This may not sound like much of a difference, but remember that we are not as interested in the depth as we are in the extent of the land that would have been exposed by these fluctuations. Since the continents typically have very shallow slopes as they fall off into the sea, a difference of 100 metres can make a huge difference in the amount of land exposed. For example, a drop in sea level of this magnitude would expose as much as 200 km of land off the west coast of India. Sri Lanka and India would have been connected by a land bridge, the Persian Gulf and the Gulf of Thailand would have been fertile river deltas, and Australia and New Guinea would have been two bulbous extremities of a single landmass. All in all, our entire coastal route would have been much different 50,000 years ago.

What the recent sea-level rise means is that, if our coastal voyagers were living primarily off resources provided by the sea, the places where they chose to live would have been those that are now under-water. The Y-chromosome pattern in Eurasia shows that our M130 coastal marker is found predominantly in the southern and eastern parts of the continent. Furthermore, M130 chromosomes in the south appear to be older than those found further north, suggesting a later migration originating in the tropics. These results, coupled with a lack of archaeological evidence for modern human occupation until after 40,000 years ago, suggest that the early coastal migrants did not stray far from the sea. Adapted to a coastal lifestyle, the surfers do not appear to have made significant colonial forays into the turf. Knowing this, it would seem more appropriate for archaeologists in search of the first Indians to be wearing scuba gear rather than pith helmets. It is likely that the earliest Upper Palaeolithic tools in the subcontinent will be found underneath thousands of years' worth of sand and coral growth.

Australian Ararat?

Laura, a small town 300 km north-west of Cairns in Queensland, is known for two things. It was once a regional headquarters for the Cape York gold-mining industry, and as such it embodied the brutality of European settlement there. Much more important to the Aborigines, though, is the fact that it is the site of the Laura Festival of Aboriginal Art and Culture, held biennially in a large field on the outskirts of town. It may seem somewhat surprising that this major international festival is held in a location that, until recently, had no paved road connecting it to the rest of the world – and one with such a legacy of colonial exploitation. Laura was chosen, though, because it is in fact the location of several sacred aboriginal sites, decorated with over 15,000 years' worth of detailed artistic depictions on the boulders surrounding the town. The art is guarded by spirits known as Quink-ans, named Timara and Imjim, who act as a kind of collective con-science. Timara is the more devilish of the two, serving to keep the population in line, while Imjim – who is characterized by a bulbous penis in his portraits – is rather more Puckish, and enjoys playing practical jokes.

The Quinkans, and their ancient pedigree, demonstrate the strong sense of connectedness that the Aborigines feel to the land where they live. Their songlines trace ancient journeys across the landscape, providing a genealogical link back to the earliest days of human existence. Like many indigenous peoples around the world, the Aborigines believe that they have always lived in their land. They cite the scientists' ever-changing estimates of the dates of human occupation, which have increased steadily over the past half-century from a few thousand years in the early 1960s to as much as 60,000 years today. As new dating methods – each with their own sources of error – have been applied to Australian prehistory, they have extended the age of human occupation there. As we'll discover, evidence for modern human occupation in Europe does not pre-date 40,000 years ago, meaning that Australians certainly have a much more ancient connection to their homeland than the Europeans who colonized their continent over the past 200 years.

The genetic results, though, clearly show that Australians – like everyone else alive today – trace their ancestry back to Africa. The Australians have an answer for this. Greg Singh, an aboriginal artist living in Cairns, suggests that the world was actually settled from Australia, explaining the genetic connection between the Oz and Africa. He claims that, as with radiocarbon dating methods giving way to thermoluminescence, so too will a reassessment of the genetic data provide evidence for the centrality of Australia in world genetic prehistory. This is clearly impossible – Africa is unequivocally where our species originated – but we could ask if the route leading to Australia, which delineates the earliest territory to be settled outside Africa, acted as a kind of prehistoric Ararat. Was the coastal route a staging post for the settlement of the rest of the world? If Africa was first, could Australia or south Asia have been the main conduit through which the rest of our journey flowed?

To find the answer to this question we will have to return to Africa, in search of the other main line of human genetic diversity.

5
Leaps and Bounds

Language is the dress of thought.
Samuel Johnson, *Lives of the English Poets*

My Y-chromosome is defined by a marker known as M173. What this means is that at some point in the past, a man – one person – had a change from an A to a C at a particular position in the nucleotide sequence of his Y-chromosome. This man could, in fact, be called M173, after the marker. All of his sons also carried this marker, marking them uniquely as his male descendants. They in turn passed it on to their sons, and over time it increased in frequency. Today, M173 is very common in western Europe, where my male ancestors come from – over 70 per cent of men in southern England have it, showing that we all have the same recent ancestor. But this is not the only marker I have – if I trace my genetic lineage back in time, I also have additional polymorphisms with names such as M9 and M89 – each one a unique change at a different position in my Y-chromosome sequence. I also have the marker M168, which places my ancestors, like those of every other Eurasian, in Africa around 50,000 years ago. The order of these markers allows me to trace the journey taken by my ancestors to the British Isles over the past 50,000 years – and exposes some fascinating relationships among people around the world. Of course, this same exercise can be repeated for every man alive today. It is a bit like deconstructing the bouillabaisse recipe passed to me by my parents and tracing back through each of the ingredient changes made by preceding generations to the ultimate source of the recipe – the original African soup.

We saw in the preceding chapter that one marker, M130, defines

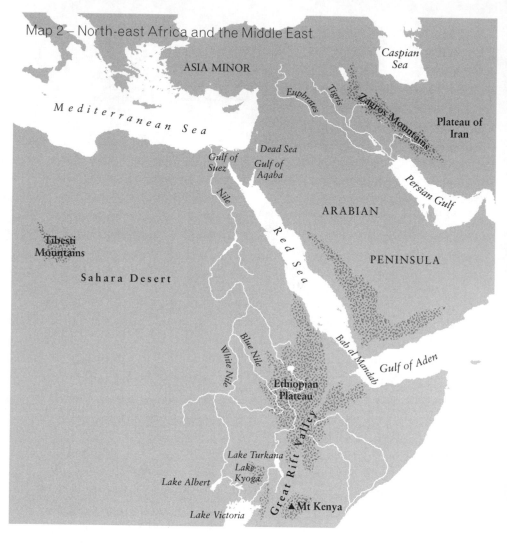

Map 2 – North-east Africa and the Middle East

the majority of men living in Australia. It traces back to Africa, sharing a common ancestor with my Y-chromosome when we reach M168 – our Eurasian Adam. The limited distribution of M130 in populations around the world reflects its coastal journey out of Africa, skirting along the south of the continent, leaving tantalizing traces of the trip along the way. But were they accompanied by men carrying M89, the next marker on my lineage? Did the journey into Eurasia begin 50,000 years ago on the south Asian coast?

Before we can answer this question, and begin to dissect the Eurasian soup recipes beginning with M89, the ancestor of most non-African Y-chromosome lineages, we need to ask a vitally important question: if – as all of the archaeological evidence suggests – modern humans were present in Africa by 150,000 years ago, why did they wait so long to leave?

Mental gymnastics

The sun is going down on the east African savannah and it is starting to get noticeably colder. You shiver, relieved that you and the other members of the hunting band managed to kill a lame gazelle. The clan will eat well tonight. When you return to camp, everyone takes a simple stone cutting tool – sharp on one edge and blunt on the other – and butchers the animal. The tool, which anthropologists today would classify as Mousterian (Middle Stone Age), is simple but effective. You make quick work of the sinew and bone and soon you are relaxing around the fire, watching the meat cook over the flames. A hyena howls in the distance and you begin to think about other things for the first time in hours.

As you mull over the day's hunt you are thankful that your luck has held out again – the animal herds do seem to be getting sparser. Of course you don't know it, but the African climate has been getting drier, and the resources to support the herds are simply not as common as they used to be. After dinner your mate brings your son to you. Although he is a strong, healthy child, he worries you because he seems so unlike the other children. For one thing, he has already learned to speak – at age two – while the other children do not do this until they are at least three. He also seems to be much better at making things than the other children in the clan, and enjoys playing games with the small pieces of stone that lie scattered around the camp. He seems much more emotional than the others, often erupting into violent temper tantrums that scare the other clan members. The strangest thing, though, is that he has begun to trace images in the dust that are similar to the animals that you bring back to camp. You find this especially frightening, and quickly rub them out when you see them. Others in the clan have noticed, though, and there have been some mutterings about his unusual behaviour.

Time passes. As your son grows up you teach him to hunt and make simple tools, but his knowledge soon surpasses yours. He seems to have a magical ability to anticipate what the animals will do, which makes him a popular member of the clan, in spite of his odd behaviour. At an early age – around fifteen – he becomes the accepted leader of your small group. Under his guidance your clan eats well and prospers. He fathers many children, and they too seem to be much cleverer than others in the group. Within a few generations all of the members of the clan can trace their ancestry to him. He becomes the 'totemic ancestor' of the group – the founding father – and everyone descended from him is by definition a member of the group. Other clans, denied the mysterious knowledge of animal behaviour and superior tool-making ability that give his clan such an advantage on the hunt, either move away or are disbanded in raids organized by the clever ones. The women are taken by the raiders and are incorporated into the clan structure, but the men are usually killed or chased away. Soon there are too many clan members to live in one small territory, and in the ensuing arguments over access to food some of the young men take their mates and set off to find new territory. The process is repeated many, many times over the next few thousand years, until essentially every man in the region traces his ancestry to that first, clever child.

What I have just described is a process that could have occurred around 60–70,000 years ago in Africa, as a single fortuitous event changed the course of human evolution. As with many historical events, it depended on the right person being in the right place at the right time – a coincidental triptych that provided the spark for a revolution. But is this necessarily the way things happened?

The short answer is that we simply don't know. The anthropological term 'Great Leap Forward', coined by Jared Diamond, was borrowed from Mao Tse-tung's 1950s plan for the industrialization of China to describe the development of radical shifts in technology at the onset of the Upper Palaeolithic, around 50–70,000 years ago. These 'killer applications', as we called them in the last chapter, marked a radical departure from the way of life that had gone before, and they deserve an explanation. What caused human behaviour to change so significantly?

Richard Klein, one of the strongest supporters of the Great Leap Forward theory, cites three significant archaeological shifts that occurred around this time. First, the tools used by humans became far more diverse and made more efficient use of stone and other materials. Second, art makes its first appearance, with a presumed leap in conceptual thought. And finally, it is around this time that humans began to exploit food resources in a far more efficient way. All-in-all, the evidence points to a major change in human behaviour. And Klein points to our DNA as the reason.

The sorts of changes we see at the onset of the Upper Palaeolithic could only have come about, he argues, if we began to communicate with each other more effectively. He infers from this that the onset of the Upper Palaeolithic marks the origin of modern language, with its rich syntax and multitude of ways to express oneself. This flowering of language skills is thought by most anthropologists to be a critical prerequisite for further social development. The development of complex social networks is almost certainly the spark that brought about the changes in Upper Palaeolithic behaviour. And this, Klein believes, happened because of a change in the way our brains are wired, set in motion by a genetic event.

We can gain some insight into what these changes may have been by looking at modern children. Swiss psychologist Jean Piaget, working in the mid-twentieth century, developed a detailed scheme for normal child development. It involved a progression from object recognition to a gradually more complex understanding of the way in which objects relate to each other. Most of the earliest stages focus on organization of real-world objects (such as bottle, rattle, or Daddy's face) into ever more complex systems through the adaptation of behaviours (when I see Daddy's face, I usually get a bottle, or sometimes a rattle). It sounds complicated, but it does seem to explain the trial-and-error way in which children learn to interact with their world. It also provides a framework for the acquisition of language skills, the most uniquely human behaviour.

Children begin to speak by 'babbling' – random sounds that roll off the tongue. This babbling phase gives way at around twelve months to actual words. Many psychologists and linguists think that children's first words, such as 'mama' and 'papa', are the easiest to learn, genetically programmed into human vocal anatomy in some way. They are found almost universally in all languages, suggesting that there may be a grain of truth in this. The American linguist Merritt Ruhlen, however, argues that the universality of these words is the evolutionary remnant of a common origin for all human languages – a trace of the original language spoken tens of thousands of years ago – rather than a programmed anatomical by-product. It is likely that both attributes play a role, with the most basic sounds having been used in the first human language *because* they are the most basic sound combinations produced by our vocal machinery.

The babbling and single words continue for another year, with a massive expansion of the child's vocabulary. The first two-word sentences begin to emerge during this process, as the child combines different words to form a clause with a new meaning. My older daughter's name is Margot, and during this phase she began to say things such as 'Margot kiss' and 'Mummy hold'. Then, around age two, a massive leap in spoken language occurs. It is at this age that most children begin to put together three words into complex sentences – 'Margot kiss Daddy', rather than simply 'Margot kiss' or 'Kiss Daddy' – with the subject-verb-object (SVO) structure, or syntax, that characterizes English and most other human languages. The structure SOV ('Margot Daddy kiss') is used by a few languages (Japanese, Korean and Tibetan among others), while VSO and VOS structures are used by around 15 per cent of languages (Welsh is an example of the former and Malagasy of the latter). The rarest structure of all is OSV, perhaps best known from the film *The Empire Strikes Back* as the language of Yoda the Jedi master: 'Sick have I become' and so on, used by only a handful of languages spoken in the Brazilian Amazon.

The important thing to glean from this syntactic diversity is that word order plays a crucial role in our understanding of a sentence. As the old saw goes, 'dog bites man' is mundane trivia, while 'man bites dog' is newsworthy.

So, the explosion of linguistic complexity in a two-year-old is a result of the mastery of syntax, and from then on it is a never-ending barrage of ever more complex sentences. The great leap forward in understanding, however, involves crossing the syntax barrier – without a mastery of this, the rest will never happen. This is what we see with chimpanzees taught to use American Sign Language, such as Kanzi the bonobo. Kanzi was able to create and understand a wide variety of two-word sentences, like an eighteen-month-old human infant, but he never mastered the complex syntax of a two-year-old's speech. The significant difference in human vs. ape communication seems to have been the creation of brain structures that allowed an understanding of syntax, and thus the communication of complex meaning.

To see why this might be, let's try another thought experiment. Imagine that you are cast away on a remote island into a tribe of people speaking a language unintelligible to you. Nothing in the language makes sense – there are no cognates with anything in your mother tongue. Your goal is to find out where you are and how to return home. How would you do it? Initially, it is likely that you would try to communicate using the skills that you developed as a young child – trial and error, focusing on nouns and verbs in isolation. Pointing to a tree, you raise your eyebrows questioningly, relying on the near universality of many human facial expressions (themselves perhaps an evolutionary remnant of a time before complex speech developed). Soon you learn enough words to develop basic sentences – 'I drink', or 'Eat now'. The final leap will be to create complex sentences that convey much more information than single nouns and verbs alone. You congratulate yourself on the achievement of two-year-old speech when you can finally say 'I go home now'. At this point, the locals seem to have a 'eureka moment', whereupon they take you to the other side of the island, to the local airstrip where you can catch a flight home.

This imaginary shipwreck scenario serves to demonstrate the utility of syntax for human communication, and gives us a good idea of why it might have been such an enormous leap forward for our early human ancestors. What it fails to do, though, is to explain what may have caused it to happen. If the intellectual chasm between humans and apes is spanned by a syntactic bridge, we need to ask why it appeared in our ancestors but not those of chimps and gorillas. Here again we obtain some help from primate behavioural research. One of the things that prevents chimps from developing complex syntax, according to Sue Savage-Rumbaugh, is limited short-term memory. To understand the meaning of a complex sentence, you must remember the beginning when you reach the end in order to integrate them. Not difficult, perhaps, for 'man bites dog', but a little tougher for a complex past-tense construction in German, where the active verb in the sentence only shows up at the end! Limited short-term memory may be the root cause of chimpanzees' minimal language skills.

The reason why our ape cousins never evolved short-term memory comparable to ours may have to do with their lifestyle. All of our simian relatives live in forests, and are at least partially arboreal. Our ancestors, on the other hand, appear to have given up a life in the trees several million years ago. Australopithecines had an upright stance, something that would only have been evolutionarily useful in a tree-free environment. The structure of the African ecosystem, with its vast savannahs in close proximity to forests, is in fact an ideal habitat for a primate making the transition from trees to the ground. And it was this leap beyond the trees that set in motion the evolutionary trajectory that would eventually lead to syntax and modern language.

Most anthropologists now accept that early hominids walked upright before they developed higher mental capabilities. As with Raymond Dart's Taung baby, the brains of the earliest human ancestors were comparable in size to those of apes, while they already showed the skeletal modifications that indicate bipedalism. Bipedalism would have conferred, in a treeless environment, the advantages of height (allowing improved vision), efficient overland movement and free hands for tool use – none of which would be terribly important if you moved primarily by climbing from branch to branch in a forest. As the saying goes, necessity is the mother of invention – and this is certainly true of evolution. But what drove us to the grasslands in the first place?

Climatic changes have periodically wreaked havoc on Africa's forests, with low rainfall reducing their area substantially several times over the past 10 million years. One particularly dry spell, between 5 and 6 million years ago, actually resulted in the disappearance of the Mediterranean, with significant knock-on effects on the African climate. During this prolonged drought some of the tree-dwelling apes may have moved to the edge of the forest to take advantage of resources offered by the grasslands. But while forest-dwelling apes are gatherers (chimpanzees occasionally kill and eat monkeys, but their diet consists primarily of fruit and insects), those who moved on to the savannah had to become hunters. This is because it is quite difficult for large primates to live on the savannah by gathering alone – plants and insects simply don't provide enough nourishment. Animals, particularly mammals, provide a high-calorie diet rich in protein. And it was the necessity of hunting and killing the mammals of the grasslands, as well as escaping the attentions of the other carnivores living there, that probably drove the development of the human brain.

If you imagine life as a chess game, then the causes and effects of brain evolution make a bit more sense. When times are good, and the environment is constant, chess can be pretty basic – perhaps even defaulting to fool's mate. If you are hungry, you find a piece of fruit or use a blade of grass to fish termites out of a hole. Simple. Life in the forest is like this, day in and day out. The reason why so many species become extinct when forests are destroyed is that they are simply unable to cope with the new environment – they are too well adapted to their local habitat. Orang-utans are gloriously suited to life in the south-east Asian rainforest, but they do not manage very well in deforested slash-and-burn fields. When times become more difficult and the environment changes, you must start to anticipate your moves in advance – and chess becomes a more challenging proposition. This is what humans thrive at, precisely because of our birth as a species in the crucible of a marginal and changing environment. In a sense, we are biologically adapted to adapt. But while other animals have complex physical adaptations, we have only our minds, and our adaptations come in the form of behavioural changes.

One of the results of having a highly adaptive mind is the development of a complex culture. Initially perhaps an extension of cooperative hunting technology, with its strong selection for intelligence and social interaction, human culture reached beyond the merely practical to encompass art, science, language and all of the other accoutrements of the 'humane' life. While we are not the first hominid to display extraordinary cultural adaptations, we are the only one to have taken them to such extremes. The Neanderthals, for instance, show evidence for group care of the sick. They also, at sites such as Teshik-Tash in present-day Uzbekistan, hint at deeper conceptualization of their place in the world, as suggested by the ritualized burial of a Neanderthal child surrounded by goat horns. But, more than any other species, it is complex culture that uniquely defines *Homo sapiens*, that makes us what we are. Without the early sparks of it, our hominid ancestors would never have ventured beyond the African forest margin into the savannah. And without having it in spades, we would never have survived what we encountered when we moved out of Africa into Eurasia, around 50,000 years ago.

Bacterial soup

When a single bacterium is placed into a nutrient-rich broth and allowed to divide to form two bacteria, then four, then eight and so on, an interesting thing happens. As we have seen, whenever DNA is copied – during reproduction – there are random mistakes known as mutations. These are the changes in soup recipe that occur naturally as a part of passing it on to the next generation. The same pattern is seen for dividing bacteria. Thus, in our rapidly propagating bacterial soup, we begin to see unique genetic lineages taking shape as a result of the small changes in their genomes. If we examine a sample of DNA sequences from the bacterial population after a few generations, we see barely any differences among them. But if we wait a few hundred generations (only a couple of days for bacteria) we see an enormous amount of variation. As with Zuckerkandl and Pauling's insight into protein evolution, the longer the population has been growing, the more variation we see. Simply put, there are more genetic differences between two bacteria chosen randomly from the older population than from the younger.

The experiment we have just performed with our bacterial soup illustrates what happens in any exponentially growing population, where we double the number of offspring each generation. Most obviously, the population increases in size rapidly – if we actually allowed the bacteria to divide without constraint for a few days, they would take over the planet. Far more important for our story, though, is the reason for this massive population explosion: every individual in the population leaves offspring. No one loses out in the evolutionary lottery – they all have bacterial babies, and their babies all have babies, and so on. This has an interesting knock-on effect on the genetic structure of the population.

If we ask how many genetic differences, on average, distinguish the bacteria that comprise the growing population, we now know that the answer depends on how long the population has been growing. In fact, there is a *distribution* of differences among the individual bacteria, rather like the bell-shaped Gaussian curve that tormented us in our mathematics classes at school. The *mean* of this distribution – the average number of differences between individuals in the sample – depends on the length of *time* that the population has been growing. If we imagine the curve as a wave, moving from left to right as it accumulates more and more differences, then the further to the right it is (in other words, the further from zero), the more mutations the population has accumulated. And like the comparisons of haemo-globin sequences from horses and gorillas, the rate at which the wave moves from left to right is predictable, because the rate at which mutations occur is constant – our molecular clock tolling in A (as well as C, G and T). Because of this, we can calculate how long the population has been growing exponentially by measuring the mean of the distribution – the midpoint of the wave. Fine, you may be saying, this may make an interesting laboratory exercise for a university genetics course, but it isn't terribly pertinent . . . unless, of course, we see the same pattern for other organisms.

Henry Harpending, an anthropologist at Pennsylvania State University, and his colleagues did precisely this analysis for the distribution of genetic differences among human mitochondrial DNA sequences and found a striking pattern. First, the distribution of differences – called the mismatch distribution – indicated quite clearly that human

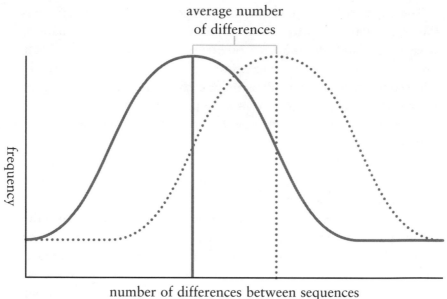

average number
of differences

frequency

number of differences between sequences

— population 1 (e.g. Europeans)   ···· population 2 (e.g. Africans)

Figure 5 Mitochondrial DNA (mtDNA) mismatch distributions of
two expanding populations. The longer the population has been
growing, the greater is the average number of sequence
differences.

populations had indeed been growing rapidly, like bacteria. This was
because the telltale wave was there in the data – a smooth, bell-shaped
curve that indicated the human species had been expanding at a great
rate. In populations of constant (or shrinking) size, the distribution
begins to deteriorate, becoming ever more saw-toothed as time goes
on owing to the uneven loss of genetic lineages – the result of genetic
drift, or perhaps selection. So, there was a clear genetic signal that
humans had expanded rapidly. The exciting result came when Harpending calculated the estimated start of the expansion: approximately
50,000 years ago, corresponding very well with our estimate of the
time at which modern humans started to migrate out of Africa, and
almost exactly with the onset of the Upper Palaeolithic.

Harpending and his colleagues examined mtDNA data collected
from twenty-five worldwide populations, and all but two of them

showed evidence for exponential growth over the past 50,000 years. The two populations with saw-toothed distributions had (on the basis of other evidence) recently been subject to drastic reductions in population size, so the analysis was clearly capable of differentiating between the two scenarios. Furthermore, the populations seemed to have expanded nearly independently of each other. Africans started the ball rolling around 60,000 years ago, followed by Asians at 50,000, and finally Europeans at 30,000 years ago. It was a stunning result. The mtDNA data agreed perfectly with archaeological evidence for the progress of Upper Palaeolithic technology: first in Africa, followed by Asia, and finally Europe – even the dates were the same. It seemed that the Great Leap Forward had left its genetic trace in our DNA, tracing the progress of the 'killer app' around the world. It also hinted at a route – but the details of the journey would have to wait until Adam's sons showed the way.

The Big Chill

When I was growing up in a city called Lubbock, in the so-called Panhandle region of Texas, we used to relate geographic distance in the form of time. The distance between Lubbock and Brownfield, a nearby town, was often given as 'around forty-five minutes', rather than 50 miles. This stems from the fact that everyone taking this journey would be driving a car, and most drivers would settle on a speed of around 60 mph – giving us a rough-and-ready conversion between time and distance.

For most of human history, distance has been expressed in a similar way. The earliest humans would have described distances in terms of the time taken to walk there. I am writing this in a house in East Anglia, near the market town of Sudbury, but if I were describing it to a Palaeolithic ancestor I might mention that it is around three days' walk from London. Similarly, our ancestors living tens of thousands of years ago would have envisioned their territories in terms of the time and effort required to traverse them. Luca Cavalli-Sforza and archaeologist Albert Ammerman have calculated that agricultural populations expanding into new territory disperse at a rate of approximately 1 km per year. Hunter-gatherers, being more mobile, can move at several times this rate. Of course, this is actual expansionary movement – the total distance walked in any year would be much more than this. But a few kilometres per year is a good estimate of the average rate at which modern-day hunter-gatherers, living in much the same way as our Upper Palaeolithic ancestors, migrate through new territory.

Based on this rate of movement, the trip from north-eastern Africa to the Bering Strait, on the opposite side of the Eurasian landmass, would have taken several thousand years. Today it is theoretically possible to make this trip in a single aeroplane flight – taking off in Djibouti (just across the Gulf of Aden from the Arabian peninsula) and landing in Provideniya, Russia, a short hop from Alaska. But around 50,000 years ago, when our ancestors began their voyage across the continent, it would have been unimaginable to make such a massive leap in one go. Rather, the journey across Eurasia would have happened imperceptibly, measured on a different time scale – one of intergenerational distances. This 'deeper' clock would have ticked away as individual bands gradually migrated into new territory, following animals, searching for water or plants, or perhaps stone for making tools. Some movement may even have been instigated by conflicts with other human groups. It was probably a combination of all of these reasons, as well as others we can't envision today. Whatever forces led to what palaeoanthropologist Chris Stringer has called the 'African Exodus', the journey must not be seen as a conscious effort to traverse the continent, but rather as a gradual expansion in range driven largely by seemingly insignificant local decisions. It is not unlike the act of squeezing toothpaste through a tube, where climate is both the stick and the carrot of the scenario. Difficulty at home forces the migration, but climatic change may lead to the appearance of new resources in distant regions. The human population is gradually forced through the geographic 'tube' by the combination of these forces, pushing and pulling over thousands of years until humans have dispersed far from their original homeland.

While this is a fair description of what motivated the earliest humans to move across Eurasia, we are interested in using the genetic data to infer the details of how it might have happened. Genetics has answered the question of *who* (Africans) and *when* (50,000 years ago), and we have some theories as to *why* (environmental change), but we now must ask *how* our ancestors of 50,000 years ago made the leap into Eurasia – and what route they would have followed. For this, we need to go back to our study of palaeoclimatology and ask what north-eastern Africa would have looked like fifty millennia ago.

The world was getting colder around 70,000 years ago, as the last ice age accelerated into a deep-freeze. This may have been the catalyst for the Great Leap Forward, favouring intelligence and complex social structures as the climate deteriorated and life became more difficult. The forests were shrinking, replaced in eastern Africa by savannah and steppe grasslands, with their wealth of large ungulates. It was on these grasslands that humans tracked and hunted, developing ever more complex tool-making and social skills. Life was incredibly active, with all effort focused on killing and gathering enough food to survive. The bell-shaped mtDNA mismatch distribution suggests that they were quite successful at this, with an expanding population even as the world turned colder and nastier.

No doubt it was the competition and difficulty of living inland, with dwindling access to both water resources and easy prey, which led some populations to live on the coast. These would have been the ancestors of the Australians, who almost certainly began to migrate out of Africa along the southern coastal route as soon as the conditions created an easy exit to the Eurasian landmass. This would have been easiest between Djibouti and present-day Yemen, a straight shot out of the Rift Valley to the endless beaches of southern Asia.

The lifestyle of these coastal people would have been relatively sedentary, tied as they were to gathering from the sea. Their days almost certainly would have been dictated by the exposure and submersion of the intertidal zone, with its rich beds of molluscs and crustaceans. Although they would probably have hunted as well, they would have been guaranteed a better return on their labours if they stayed near the coast. As we saw in the previous chapter, the genetic and archaeological data bear this out, suggesting that they did not stray very far inland at this early stage. The interior was left to the more active hunters, who would have had to move great distances to obtain the resources they needed to survive – animals, plants and water. They are the ones who made the leap into the unknown beyond the coast, into the wilds of interior Eurasia.

One of the apparent conundrums of biology is that the more temperate parts of the world actually contain the largest animals. In ecology there is an observation known as Bergmann's rule, which states that body size increases with latitude. While this isn't strictly true for every species, it is a good generalization. The woolly mammoths, largest land mammals of the past few hundred thousand years, lived in the tundra regions of far northern Eurasia and America. In the sea, there is actually more biological material in the colder parts of the planet than there is in the warmer. In spite of the incredible diversity found on a coral reef, the total mass of organisms is significantly less than that found in more polar regions. The polar oceans, for instance, contain the world's densest concentrations of plankton. These tiny plants and animals support the largest animals on earth, the filter-feeding baleen whales which, over time, have become almost completely dependent on this unusual food source, the remnants of their terrestrial lives tens of millions of years ago being nearly invisible today.

Similarly, the tropical rainforest contains a huge number of species, but the size – and density – of any particular species is quite low. Furthermore, because all of the nutrients are tied up in organisms, the soil actually contains very little in the way of minerals and organic matter. In reality, overgrown bushes do not typically clog the ground in a mature tropical forest, despite the Hollywood cliché of machete-wielding explorers. The tragedy of deforestation is that it is all too easy, in the space of a few years, to reduce a teeming ecosystem to a lifeless desert. The tropical environment is poised precariously on the edge of fecundity and death, extremely susceptible to relatively trivial disturbances.

The temperate parts of the planet, on the other hand, are endowed with rather more resilience. While the species diversity is a small fraction of that seen in the rainforest, the organisms living there are better able to withstand drastic upheaval. This is primarily due to the vicissitudes of life in the temperate zone. Tropical climatic stability has nurtured the evolution of species over tens of millions of years in virtually unchanged conditions (save for variations in geographic

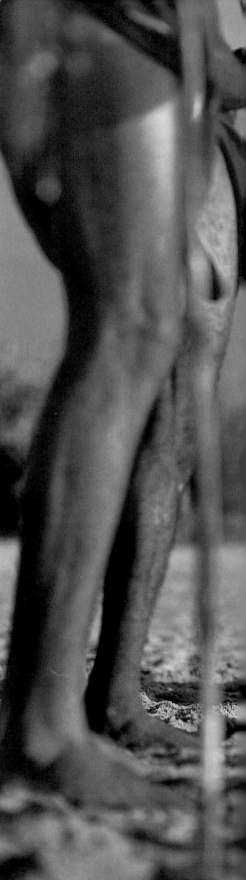

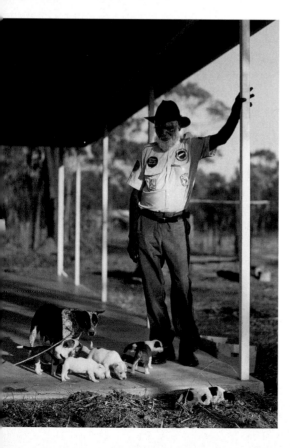

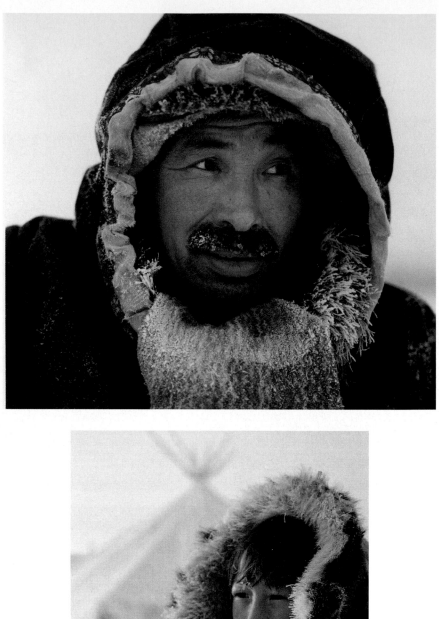

range). On the other hand, vast tracts of the Eurasian landmass have been periodically covered in ice or reduced to deserts during the same time period. This long-term cycle actually mirrors the annual variation in weather that produces the temperate zone's seasons: the dry heat of a Mongolian summer yields to icy winter storms in the space of a few months. Because of the enormous environmental variation seen there, animals living in the temperate zone have had to rely on two crucial adaptations to keep themselves alive: investment and migration.

In the same way that you or I might choose to forgo the instant gratification of spending every penny we earn in a non-stop shopping binge, with an eye to using the money saved to see us through difficult times or old age, so animals that are used to encountering difficulty set aside some of their resources during times of plenty. It is not a conscious decision, but rather an evolved instinctual behaviour – an adaptation to life on a meteorological seesaw. Every spring and summer, for instance, the arctic tundra explodes into an orgy of growth and reproduction. Plants flower, pushing shoots above the permafrost for the first time in nearly ten months. Mosquitoes engulf everything that moves in a buzzing, bloodsucking cloud and the mammals that live in the Arctic – such as reindeer and walrus – give birth to their young. During this benign period, when temperatures can soar to nearly 100°C above their winter lows, you would be forgiven for thinking that the far north is one of the most productive places on earth – a teeming mass of life, hell-bent on one last bang before winter sets in again and everything dies. However, there is a method in the madness of the creatures living in the far north. It is during this time that every species in the Arctic is preparing for the end of the party, which will come like clockwork in early September when the temperature drops below freezing once again. No tropical mammal would ever evolve the behaviour of building up fat reserves to prepare for times of famine, but most temperate species do this as a matter of course. During the Arctic summer, reindeer add as much as a third to their body weight, storing resources for the long, dark winter. This allows them to survive the period of dearth that comprises 70 per cent of the year. It also makes them tempting targets for carnivores.

Humans, as they adapted to life on the plains of east Africa, would have become more and more adept at hunting the large mammal

species that lived there. These include several species of antelope, an animal that has been called the 'takeaway pizza' of the Upper Palaeolithic. One of the behavioural changes that may have taken place around the time of the onset of the Upper Palaeolithic is the specialization of human populations on particular prey species, with presumed adaptations in hunting methods and weapons. The techniques you use to take down a gazelle, for instance, are quite different from those you might use to kill a mammoth or a rhinoceros. Specialization would have allowed efficient use of the animal resources in a region – but it would also have led to more movement, as the herds were depleted in one region and it became necessary to move on to find others in distant places. Seasonal hunting also appears to have made its appearance around this time, with evidence that early human populations followed herds of grazing animals – particularly antelope – from summer pastures in the hills surrounding the Mediterranean and the Red Sea, down to the warmer coastal regions in the winter. It was this gradual ebb and flow of animals, over hundreds or thousands of years, that may have brought fully modern humans and their Upper Palaeolithic toolkits into the Middle East around 45,000 years ago.

Modern humans had been present in the Levant (the eastern region of the Mediterranean) since at least 110,000 years ago, but the population was never extensive and was limited to a few sites. During this early phase of the last ice age, the eastern Mediterranean was effectively an extension of northern Africa, with similar climatic conditions and animals. The cave sites of Qafzeh and Skuhl in present-day Israel contained typically Ethiopian animals during the time when modern humans occupied them. Then, during the period between 80,000 and 50,000 years ago, modern humans abruptly disappear from these sites. In some cases they were replaced by Neanderthals, with their thick skulls and robust skeletons. This gives us a clue as to what was happening in the Levant during this time.

The climate was getting much colder after 80,000 years ago, and temperatures around the eastern Mediterranean were plummeting. It is likely that the average global temperature during this time dropped by around 10°C, with massive knock-on effects on plant and animal distributions. The early modern humans who had migrated out of Africa via Egypt and the Levant during the wetter and warmer times

found that they were unable to rely on animals they had hunted for thousands of years. They may have died off, or perhaps they simply migrated back to Africa, but they do not seem to have made any further headway in their conquest of the Eurasian interior. These early modern humans are probably best viewed as a tentative stab at the world beyond Africa – one that simply did not make it any further.

Then, around 45,000 years ago, modern humans appear again in the Levant. This time, though, there was a critical difference. While the humans of 40,000 years earlier had used tools very similar to those of their Neanderthal contemporaries, the latest invaders carried with them the 'killer app'. These people were the recent inheritors of the Great Leap Forward, with its advanced technology and complex culture in tow. Their Upper Palaeolithic tools and cooperative hunting behaviour – as evidenced by seasonal migrations and prey specialization – gave them an edge that the earlier moderns had lacked. Once they entered the scene, the path was open to the rest of the continent.

The route they followed in their blitz across Eurasia is revealed by the genetic patterns, so for the next part of the journey we'll need to leave aside stones and bones and return to our DNA excavation.

6
The Main Line

Now here, you see, it takes all the running you can do, to keep in the same place. If you want to get somewhere else, you must run at least twice as fast as that!
Lewis Carroll, *Through the Looking Glass*

As I mentioned at the beginning of the last chapter, my lineage of Y-chromosome markers coalesces back to a DNA polymorphism known as M168, the ancestor of everyone living outside of Africa. M168 unites me with the Australian coastal migrants, tracing us both back to Africa around 50,000 years ago. This places all non-Africans in that continent immediately after the earliest archaeological evidence for the Great Leap Forward, and suggests a causal relationship between this ancient cultural revolution and the migration of modern humans out of Africa. The people who stayed in Africa, as well as the ones who left, would have been fully modern in every respect – technologically, culturally and artistically. The mitochondrial DNA results suggest that a massive expansion in human populations began around this time, consistent with the range expansion we see in the archaeological record. The Y and mtDNA data hint at two routes, one of which would have followed the coast to Australia around 50–60,000 years ago. What about the other, which accounts for the majority of people in the world today?

Before we begin to trace the order of the additional markers on my lineage, and their significance to our story, we need to clarify what the order actually signifies. There are two issues to be considered here, and both involve timing. The first is what we might call relative dating. To understand this, it is worth revisiting our hypothetical kitchen. Like the maternal and paternal soup recipes, the genetic recipes we have all inherited contain a combination of ingredients, or markers, that distinguish them from everyone else's soup. In order to establish the order in which the ingredients were modified, we need to compare many different recipes before we begin to see patterns. So let's do a bit of genetic cooking.

Imagine having an international potluck supper, where everyone invited is asked to bring a soup that is specific to his or her own country. In our kitchen we have several dozen bowls of soup sitting on the table. Each has a slightly different recipe, but they all come from the same source. How do we know this? Because each recipe uses as its basic ingredient impala – a species of antelope that occurs naturally only in Africa. It is extremely difficult to obtain impala meat in many parts of the world, but it is the cornerstone of all the soup recipes and it must be included.

As we taste the soups, we begin to detect another pattern. Some contain black pepper, while others contain salt. These are the two main soup categories, and if you have one you don't have the other. There are many additional variants among the salt recipes – some with fish, others with barley, a few with unusual spices you can't identify – but they are all united by the presence of salt. Similarly, the black pepper recipes have a huge range of additional ingredients – thyme, berries, pork, nuts – but they all contain black pepper.

In this recipe game, we will make use of Ockham's insight into historical change to infer the order in which the ingredients were added. If we assume that the addition of ingredients occurs at a regular rate, and there is no loss or substitution of ingredients once they are added, then the most common ingredients have usually been added the earliest. This is because this order minimizes the total number of changes required to explain the soup recipes. For example, if we were to sample five soups from the ones sitting on the table, we might find the following recipes:

- impala, mustard, black pepper, cheese, oregano
- impala, salt, loganberries, peanuts, chilli peppers
- impala, mustard, black pepper, clams, basil
- impala, black pepper, crab, juniper berries
- impala, salt, thyme, parsley, pork

Map 3 – Eurasia

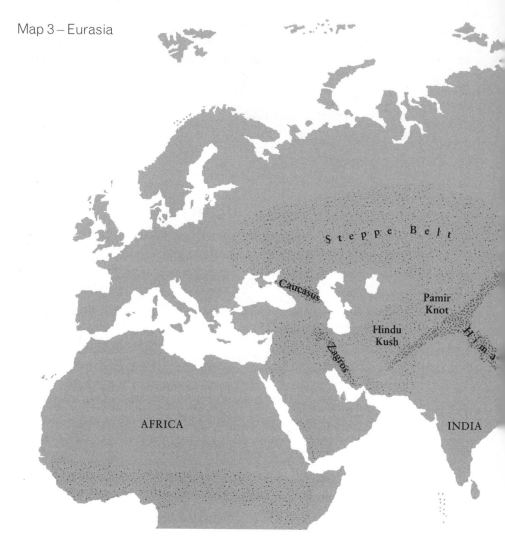

What can we say about the order in which the ingredients were added? Well, the first pattern is that all of the soups contain impala. This means that the most likely explanation is that the original soup also contained impala – much more likely than if all of the cooks had independently decided to add impala at some point in the past. Remember that this ingredient is extremely difficult to obtain in most parts of the world. The next obvious pattern is the one we noted before: some recipes contain salt while others contain pepper. By the same reasoning that placed impala first in the ingredient list, because it minimizes the number of independent, identical ingredient changes, salt or pepper define the next addition to the recipes. After that, we see that mustard unites two of the recipes, making it the next addition after black pepper. So, we have derived an order for the addition of ingredients by our ancestors: impala, followed by salt/pepper, followed by (on the pepper lineage) mustard. Order emerges out of chaos.

In the case of our soups, it may seem that we could have chosen to ignore the possibility that the same ingredient could have been added independently. Why couldn't mustard also show up in the salt recipes, for instance? While this may be possible for common ingredients, we should actually imagine the soup components as rare, home-grown varieties produced in small batches, only available in tiny local shops. In this case, it is virtually impossible for a cook from Mexico and one from Namibia to use the same type of mustard – our sophisticated palates would be able to tell the difference. Adding the same ingredient independently is almost impossible.

This sort of puzzle-solving is exactly what we have done for the markers that define our genetic lineages. If M168 defines a marker common to all non-African populations, then in our genetic recipe it is the equivalent of impala – the marker that unites everyone outside of Africa. If the lineage then splits into salt and pepper, we can imagine M130 – our Australian marker – as genetic salt, while pepper is represented by another marker, known as M89. Because of the order in which the ingredients have been added, we can infer that M130 and M89 are approximately the same age. Since we know that we were in Australia between 50,000 and 60,000 years ago, and M130 has not been found in Africa, we can set this as the upper limit for the age of these markers – it is likely that they arose at or after this time. The archaeology gives us an independent way of assessing how old they are. But what if we wanted to guess at the age using only genetic data? Could we do it?

The answer is yes, and this brings us to the other dating method – counterpart to the relative dating we used to assign the order of the ingredients. Like the isotopic dating methods discussed in Chapter 4 – particularly the ones with the intimidating names – the possibility of error is high for absolute genetic dating methods, because there are quite a few assumptions involved in the way the dates are calculated. Nonetheless, they provide us with an assessment of the age of markers – and therefore people – that is independent of the archaeological record. To see how this works, we will use our soup recipes to try to figure out the absolute ages of the ingredients – in other words, the time in the past when the ingredients were first added to the recipe.

The first rule for absolute dating, as mentioned above, is that the ingredients are added at a regular rate. The second is that once an ingredient is added, it becomes a permanent part of the recipe – there is no way to remove it later if you don't like it. From these two rules it's easy to predict that, over time, the soup recipes should become more and more complicated. The longer they have been accumulating ingredients, the more culinary diversity we should see. And because the ingredients have been added by particular people living in the past, they are like a culinary signature of our ancestors. They mark not simply the ingredients, but the people who added them. So, by dating the ingredients, we actually date the cooks who passed on our recipes.

Let's assume that the ingredients are added at a rate of one every ten generations. Most people are happy to cook the same soup their parents did, but some finicky person pops up every ten generations or so who has to change the recipe in a minor way in order to 'improve' it. We can use this to estimate how many generations ago our first impala soup was prepared. There are four additional ingredients in each of the soup recipes shown above, so we have been accumulating changes for around forty generations (4 × 10). If we assume that there are, on average, twenty-five years in each generation (the average age of parents when they have children), this gives us a time of 1,000 years that the recipes have been accumulating changes. Therefore, the person who started cooking soup with impala also lived about 1,000 years ago. We can even, by looking at where the ingredient occurs, guess at where this person may have lived. If we assume that newly added ingredients are chosen locally, then where would that person most likely have been living in order to choose impala? Since impala are an African species, Africa is the most likely place.

So, by looking at the soups and making a few assumptions about the way they change, we've been able to do two things. We have derived an order for the addition of the ingredients, and we've been able to estimate the time and place when the ingredients were added. In other words, we have used a bit of tasting and mathematics to tell us the *who, where* and *when* of soup history. Rather amazing, really – to be able to say so much from a few tastes.

In the same way that soup tasting can give us a glimpse of the culinary past, so too can genetic 'tasting' – which we call sampling – tell us about the human past. By inferring relative and absolute dates, and asking where the most likely origin would have been, we can actually trace ancient genetic migrations around the world. The first stop is at the edge of an ebbing Mediterranean wave, just before the world dried out and trapped a few people in the Middle East, around 45,000 years ago.

Continental bollards

As we have seen, the Middle East has always been an extension of north-eastern Africa, to both grazing animals and the humans that hunted them. This had been the case millions of years before, when *Homo erectus* moved into the Caucasus via the Levant soon after he appeared in Africa. Between the hominid homeland in the Rift Valley and the benign Mediterranean climate, however, lies the eastern edge of the Sahara Desert. This gives a clue about the time and the route that we can use to test our genetic estimates.

Major geographical features – seas, deserts and mountains – have always served as barriers to the dispersal of living organisms. The unique flora and fauna of Australasia, for instance, have been maintained by the presence of an unbroken water barrier between this continent and the rest of the world. Similarly, mountains can act as barriers, incongruous pieces of arctic real estate that serve to deter movement. In a way, geographical barriers are like bollards – those raised reflective markers that serve to guide automobile traffic.

While seas and mountains are (at least on the time scale of human evolution) huge barriers to movement, deserts are much more fluid. As we have seen with the forests and savannahs of Africa, deserts are interchangeable with other ecosystems. If the rainfall drops below a certain level, desertification can happen nearly overnight. Similarly, increases in rainfall can reclaim fertile land from the sands just as suddenly. Because of this, deserts should actually be seen as ebbing and flowing ecosystems, extending their range when the climate is dry and losing it when moisture is more plentiful – like waves lapping at the edges of the other ecosystems. Paraphrasing that old saying about British weather, if you don't like the desert, simply wait a few hundred years and it will change.

The largest desert in the world is found in Africa: the Sahara. It evokes images of rolling dunes, camels, oases, date palms and extreme heat – the name is almost synonymous with desert. It has served as an extraordinary barrier to human movement throughout recorded history, to such an extent that Africa is divided by geographers in two zones: Saharan and sub-Saharan. The Saharan region has historically been closer to the Mediterranean world, since human settlement was limited to a narrow strip along the coast. The sub-Saharan zone, well beyond the Pharaonic sixth cataract of the Nile, was a distant and mysterious place, isolated by a 2,000-km wide strip of sand and heat. Clearly a significant barrier.

The Sahara has not always been like this, however. During the early stages of modern human development, it was a relatively moist place, with a significant human presence. Middle Palaeolithic sites dating to 80–100,000 years ago have been found throughout, and it is only with the acceleration of the last ice age after 80,000 years ago that humans disappear from the Sahara. There appears to have been a short 'spike' of elevated temperatures (and thus increased rainfall) around 50,000 years ago, when the northern hemisphere warmed slightly for a few thousand years, but the general trend from 70,000 years ago is one of lower and lower temperatures. In the case of Africa, this meant drier conditions and an expanding Sahara. We know this because of an increase in sand in the sediments from the Mediterranean during this time, as well as the disappearance of savannah species from the desert itself.

The first Upper Palaeolithic humans may have reached the Middle East during the relatively warm and moist conditions around 50,000 years ago, when the eastern Sahara was in retreat and a gateway opened along the Red Sea. Perhaps they migrated down the Nile to the Mediterranean, then spread eastward across the Sinai peninsula. Alternatively, early human populations may have moved across the strait of Bab al Mandab into southern Arabia, a short hop of 20 km or so. Once there, the relatively moist conditions along the coastal mountain range of western Arabia – which served to scoop moisture from the prevailing westerly winds coming off the Red Sea – may have created savannah-like hunting conditions for these Upper Palaeolithic people. Even today there is a narrow strip of steppe extending as far north as the city of Medina in Saudi Arabia, unique in the harsh environment that defines most of the Arabian peninsula. In the past, this tenuous steppe environment may have been joined with its eco-logical equivalent extending southward from the Gulf of Aqaba in Jordan, effectively opening a door to the interior of Eurasia.

William Calvin, a neurobiologist who has written extensively on climate and early human evolution, has compared the Sahara to a kind of hominid 'pump'. During wetter periods, the Sahara would have sustained human populations, perhaps focused around oases or rivers, or limited to zones that received moisture from prevailing winds. As the conditions turned drier, the Sahara would have returned to uninhabitable desert, forcing human emigration. Calvin suggests that the climatological downturn after 50,000 years ago may have been the impetus for the migration of Upper Palaeolithic humans out of northern Africa and into the Middle East.

However the earliest Upper Palaeolithic moderns reached the Levant, it is clear that the deteriorating climate after 45,000 years ago effectively locked them into their new home. The Sahara would have been at its driest between 40,000 and 20,000 years ago, and it is likely that any previously inhabitable areas there would have been engulfed by desert during this time. Modern humans were trapped in a new continent.

The genetic pattern bears this out, and provides the next clue on our journey. M89, the marker that occurred immediately after M168 on our main line into Eurasia, has been dated using the absolute method detailed above to around 40,000 years ago. Due to possible errors in the assumptions that go into the calculation, particularly in determining the rate at which new mutations occur, this estimate actually encompasses a range between 30,000 and 50,000 years, and it is likely (given the climatic data) that it appeared at the earlier end of this range, perhaps 45,000–50,000 years ago. This is because it serves to unite populations living in north-eastern Africa – Ethiopia and Sudan in particular – with the populations of the Levant. The shutting of the Saharan gate after these M89-bearing populations were allowed through is suggested by the low frequency in north-eastern Africa of Eurasian markers that occurred *later* on the M89 lineage. If Africa and the Levant had been part of a continuous range occupied by humans throughout the Upper Palaeolithic, we would expect to see a relatively homogeneous distribution of markers throughout. In fact, it seems that the emigration of populations bearing M89, which we can call a Middle Eastern marker, signified the last substantial Upper Palaeolithic exchange between sub-Saharan Africa and Eurasia. The world had been divided into African and Eurasian, and it was to be tens of thousands of years until significant exchange was to take place again.

The presence of M89 in both north-eastern Africa and the Middle East, and the age of the Upper Palaeolithic archaeological sites in the Levant, helps us to answer the question of whether Eurasia was settled in a single southern coastal emigration from Africa. M130 chromosomes are not found in Africa, suggesting that this coastal marker arose on an M168 chromosome *en route* to Australia. Conversely, M89 Y-chromosomes are not found in Australia or south-east Asia – but they appear at fairly high frequency in north-eastern Africa. The implication is that M89 appeared slightly later than M130, in a population that stayed behind in Africa after the coastal migrants left for Australia. It was these people, *sans* M130 chromosomes, who first colonized the Middle East. There is archaeological evidence for a modern human presence in the Levant from around 45,000 years ago, consistent with the arrival of modern humans from somewhere else. North-eastern Africa is the only nearby location with archaeological

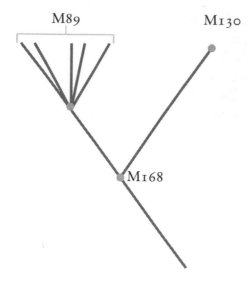

Figure 6 M89 defines the main Y-chromosome lineage in non-Africans.

sites dating from around the same time – and, crucially, the same genetic markers we see in the Levant. Thus, the genetic and archaeological patterns tell us that there was a second migration from Africa into the Middle East.

Once our Upper Palaeolithic migrants had arrived in the Levant, the road into the heart of Eurasia was open. There was a continuous highway of steppe – not unlike African savannah in terms of its species composition – that stretched from the Gulf of Aqaba to northern Iran, and beyond into central Asia and Mongolia. The hurdle of the Sahara having been overcome, the subsequent dispersal of these fully modern humans would have been limited only by their own wanderlust. They had all of the intellectual building blocks that would enable them to conquer the continent, and the process began with gradual migrations along this Steppe Highway, the continental equivalent of the southern Coastal Highway.

At this time, game would have been plentiful. The large, grazing mammals of the steppe zone – particularly antelope and bovids, the ancestors of the domestic cow – would have been easy prey for early humans, and they gradually expanded their range as their numbers grew. Moving northward and westward, some may have entered the Balkans early on – the first modern humans in Europe. The numbers would not have been great, though, since it was far easier to stay within the bounds of the steppe zone to which they had become so well adapted. The mountains and temperate forests of the Balkan peninsula would have seemed rather alien to early Upper Palaeolithic people, and the genetic data bears this out. Very few Europeans trace their ancestry directly to the Levant of 45,000 years ago, as attested to by the Y-chromosome results. Our canonical Levantine Upper Palaeolithic lineage, M89, is found at frequencies of only a few per cent in western Europe. It may have been these few Middle Eastern immigrants who introduced the earliest signs of the Upper Palaeolithic to Europe, a culture known as the Chattelperronian, but they did not leave a lasting trace. The true conquest of Europe, and the demise of the Mousterian, would have to wait for a later wave of immigration – people with a few more ingredients in their genetic soup.

Eastward ho!

The main body of Upper Palaeolithic people began to disperse eastward. As with other early human migrations, it almost certainly wasn't a conscious effort to move from one place to another. Rather, it seems that the continuous belt of steppe stretching across Eurasia provided an easy means of dispersal, gradually following game further and further afield. It was during this time that another marker appeared on the M89 lineage, given the name M9. It was the descendants of M9, a man born perhaps 40,000 years ago on the plains of Iran or southern central Asia, who were to expand their range to the ends of the earth over the next 30,000 years. We will call the people carrying M9 the Eurasian clan.

As the steppe hunters migrated eastward, carrying Eurasian lineages into the interior of the continent, they encountered the most significant geographical bollards so far. These were the great mountain ranges that define the southern central Asian highlands – the Hindu Kush running west to east, the Himalayas running north-west to south-east and the Tien Shan running south-west to north-east. The three ranges

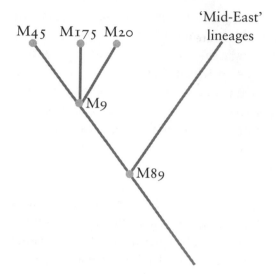

M45   M175   M20   'Mid-East'
                        lineages

Figure 7 Descendant lineages of M89 characterizing the main geographic regions in Eurasia.

meet in the centre, at the so-called Pamir Knot in present-day Tajikistan, and each radiates off like a spoke in a wheel.

The first humans to see them must have been absolutely awe-inspired. Although they had encountered the Zagros range in western Iran, it was a permeable barrier, with numerous valleys and low passes that would have allowed easy movement. The Zagros themselves actually would have been part of the geographic range of the prey species hunted by Upper Palaeolithic people, with the herds migrating into higher pastures during the summer and descending to the surrounding plains in the winter. The high mountains of central Asia were a different beast altogether. Each of the ranges has peaks that soar to 5,000 metres or higher (in the case of the Tien Shan and Himalayas, over 7,000 metres), and the radiating high-altitude ridges would have been formidable barriers to movement. Remember that the world was in the grip of the last ice age, and temperatures would have been even more extreme than today. It was because of these mountains that our Eurasian migrants would have been split into two groups – one moving to the north of the Hindu Kush, the other to the south, into Pakistan and the Indian subcontinent. How do we know this? The Y-chromosome again traces the route.

Those who headed north, toward central Asia, had additional mutations on their Eurasian lineage that we will trace below. The Upper Palaeolithic people who headed south, though, had an unrelated mutation on their Y-chromosome known as M20. It is not found at appreciable frequencies outside of India – perhaps 1–2 per cent in some Middle Eastern populations. In the subcontinent, though, around 50 per cent of the men in southern India have M20. This suggests that it marks the earliest significant settlement of India, forming a uniquely Indian genetic substratum – which we can call the Indian clan – that pre-dates later migrations from the north. The ancestors of the Indian clan, who moved into southern India around 30,000 years ago, would have encountered the earlier coastal migrants still living there. From the genetic pattern, it seems likely that any admixture with them was not reciprocal: as we saw in Chapter 4, mitochondrial DNA retains strong evidence of the coastal migrants in the form of haplogroup M, while the Y-chromosome primarily shows evidence of later migrants from the north. Thinking back to the scenario we imagined for the birth of the Upper Palaeolithic in Africa, this is the pattern we would expect to see if the invaders took *wives* from the coastal population, but the coastal *men* were largely driven away, killed, or simply not given the chance to reproduce. The result would be the widespread introduction of M mtDNA lineages into the Indian population, while the Coastal Y-chromosome lineages would not be nearly as common – precisely the pattern we see. Today, the frequency of the Coastal marker is only around 5 per cent in southern India, and it falls in frequency as we move northward. This pattern suggests that the contribution from the coastal populations was minimal, at least on the male side. The contrast between the two types of data gives us a glimpse of the behaviour of these first Indians, and hints at a cultural pattern we will explore in more detail in Chapter 8.

The migrating Eurasian masses were not only shunted down into India, of course – some of them also migrated to the north of the Hindu Kush, into the heart of central Asia. The Tien Shan would have been an even more formidable barrier than the Hindu Kush, keeping the Upper Palaeolithic hunters out of western China. It is around this time that another mutation occurred on the Eurasian lineage. It was known as M45, and it will help us to trace two very important later migrations. Using absolute dating methods, we can infer that the M45 mutation occurred approximately 35,000 years ago in central Asia. Today, M45 is found only in central Asians and those who trace their ancestry to this region – thus, it defines a central Asian clan. Descendants of the central Asian clan occur only sporadically in the Middle East and East Asia, and at somewhat higher frequency in India, where the clan appears to have migrated much later (as revealed by the presence of additional mutations). The 'ancestral' form – the deepest split in the genealogy of Y-chromosomes from the central Asian clan – is found only in central Asia. This allows us to pinpoint the location of what is effectively a 'regional Adam', in much the same way that we identified our African Adam as being an ancestor of the San Bushmen. The deepest branches in the M45 genealogy are found today only in central Asia – not India, or Europe, or east Asia. Thus, M45 arose in central Asia.

The limited distribution of the oldest descendants of the central Asian clan suggests that the population where it arose was isolated from people living in the surrounding parts of the continent. While the Hindu Kush provides a ready explanation for why there was no easy migratory path to India, it is not clear why this population had no contact with groups living in the Middle East. After all, our Eurasian clan had migrated into central Asia along this route – why couldn't the central Asian clan make the return trip? The inference is that another bollard had entered the story, and given that it hadn't been an insuperable barrier several thousand years before when the central Asia clan's ancestors first migrated to the heart of the continent, it was likely to have appeared after that first migration.

Today, the Dasht-e Kavir and Dasht-e Lut deserts of central Iran are scorched, parched wastelands. The tiny population living there ekes out a meagre living using a highly developed system of agriculture, complete with miles of underground irrigation channels known as *ghanats* that have been in use for thousands of years. During the heat of the day the residents of cities such as Yazd retire to subterranean chambers cooled by wind channelled down long pipes, creating a haunting wail that can be heard from miles away. It is inconceivable that anyone could survive for long in this harsh climate without such a well-adapted lifestyle. Hunting and gathering would be impossible – at least today. Similarly the Karakum and Kyzylkum deserts of central Asia are harsh, desolate places with very few inhabitants apart from a few nomadic shepherds.

There are, however, two belts of continuous steppe across the deserts of central Iran, one to the north of the deserts, near the Caspian, and one to the south, near the Arabian Gulf. When the world was in the midst of its climatic schizophrenia around 40,000 years ago, it is likely that the steppelands and deserts of Iran and central Asia went through periods when the amount of moisture in the atmosphere would have been similar to, or perhaps greater than, today. This could have been aided by changes in the prevailing winds, bringing moisture in off the Arabian Sea. During these relatively wet periods, which may have been brief, humans would have been able to migrate fairly easily across the Iranian plateau and into central Asia – again, the prey and hunting methods would be virtually identical throughout the entire journey. We know that they did so because of the genetic trail they left in their descendants, which traces a direct path from the Levant to central Asia.

Once the ice age reached a threshold temperature, though, there was a significant decrease in precipitation and humidity as evaporation stalled and water became frozen into the expanding ice sheets of the far north. This seems to have happened between 40,000 and 20,000 years ago, and it resulted in the creation of a new desert bollard on our route. The continent was now split into northern, southern and western populations, all headed into the coldest part of the ice age. The people living in India and the Levant had the benefit of the sea, which served to mitigate the effects of the increasingly cold and arid conditions. Those trapped north of the Hindu Kush, however, had to adapt to the increasingly harsh lifestyle of the Eurasian steppes – or die.

It is likely that these early central Asians would have stayed in the relatively warm environs of the southern steppes had encroaching desertification not forced them on. Some stayed behind, retreating into the foothills of the Hindu Kush where the water supply from glacial melting, and the number of animals, were sufficient for survival. Most, though, appear to have followed the migrating herds of game to the north – into the face of the storm, as it were. It is likely that they first reached Siberia during the early part of this period, around 40,000 years ago, when Upper Palaeolithic tools make their appearance in the Altai Mountains. The conditions would have been unimaginably different from those their ancestors had left behind in Africa 10,000 years before. Winter temperatures dropped to –40°C or lower, and much of their time would have been spent hunting for food and keeping warm. But the animals they hunted would have made the difficulties worthwhile.

We saw earlier that one of the defining features of species living at high latitudes is their great size – Bergmann's rule. The reason is that large animals have less surface area relative to their volume than small ones, and heat is lost through the surface. Shrews must eat constantly to maintain their hyperactive metabolism, in part because their tiny size makes it extremely difficult for them to retain heat. In cold environments, then, there is selection for large animals with slower metabolisms (since the food resources are not as plentiful as they are in warmer regions) – big, lumbering beasts that aren't particularly clever. This is how natural selection created animals such as the woolly mammoth.

The first Upper Palaeolithic people to encounter a mammoth, probably in southern Siberia or central Asia, must have been more than a little intimidated. While the traditional prey of these consummate hunters would have been perhaps two or three times the size of a human, mammoths were the size of small buses, with intimidating tusks and thick fur. As they watched these odd giants, though, they would have discovered that the mammoth's great size also made it slow and ungainly. Given the right hunting technique, and the right tools, it was possible to kill them. And once this was done, the meat from one animal would feed a clan for weeks, so it was worth making the effort.

It is also likely that dead mammoth carcasses were scavenged by Upper Palaeolithic humans. On the basis of animal remains at Upper Palaeolithic sites in southern Africa, anthropologist Lewis Binford has suggested that scavenging formed a substantial part of the diet of early human populations. While the relative levels of scavenging and hunting practised by our early ancestors have been debated by scientists, it is likely that at least some scavenging took place – as is the case in modern hunter-gatherer groups. And with the great mass of meat on a mammoth carcass, they would have been prime scavenging targets for the first Eurasians.

The Eurasian interior was a fairly brutal school for our ancestors. Advanced problem-solving skills would have been critical to their survival, which helps us to understand why it was only after the Great Leap Forward in intellectual capacity that humans were ready to colonize most of the world. During their sojourn on the steppes, modern humans developed highly specialized toolkits, including bone needles that allowed them to sew together animal skins into clothing that provided warmth at temperatures not unlike those on the moon, but still allowed the mobility necessary to hunt game such as reindeer and mammoth successfully. They had to venture beyond sheltering hills and caves, out on to the icy open steppe and tundra, necessitating the development of portable shelters. Their migrations would have taken them far beyond ready sources of the fine-grained stone they used to make weapons, so they had to become more economical in their tool-making. This led them to develop microliths, small stone points (such as arrowheads) that were hafted on to wooden shafts and used as weapons.

The problem-solving intelligence that would have allowed Upper Palaeolithic people to live in the harsh northern Eurasian steppes and hunt enormous game illustrates something that could be called the 'will to kill'. Survival depended on finding sufficient food resources, whatever the obstacles – and the steppes were a veritable meat locker. It was the necessity of obtaining food that led them into the freezer, but it would take them well beyond central Asia. The Steppe Highway gave them a straight shot to the extreme ends of the continent, and once they had adapted to the harsh conditions a new world lay open to them.

Chopsticks

The genetic composition of these first Siberians was a mixture of both central Asian *and* ancestral Eurasian clan lineages. While M45 is the marker that we use to infer the migrations of the early central Asian steppe hunters, there were still many men alive who did not have Y-chromosomes marked with M45 – they would have had unmarked Eurasian M9 Y-chromosomes. This is because new markers do not immediately increase in frequency to the point where all other markers – such as the ancestral M9 lineage – are lost. All of the Y-chromosome markers we study originated in a single man at some point in the past, so their original frequency was one (that individual) divided by the total number of men in the population – a very low frequency in all but the smallest groups. Over time, they become more common primarily due to the effect of genetic drift – the random changes in frequency that characterize all human populations. Thus the earliest people to colonize southern Siberia would have had members of both the central Asian M45 and the older Eurasian M9 clans, although drift appears to have caused them to lose most of their ancestral Middle Eastern chromosomes by this point.

As with the Eurasians who entered India on the other side of the Hindu Kush, some of these Eurasian clan members would have migrated to the north and east, guided in their journey by the Tien Shan mountains. Some of them, perhaps taking advantage of the so-called 'Dzhungarian Gap' used thousands of years later by Genghis Khan to invade central Asia, made it into present-day China. It is likely that the majority were migrants along the Steppe Highway further to the north, avoiding the harsh deserts of western China by detouring through southern Siberia. However, make it they did. We know this because they left descendants from another Y-chromosome marker that is almost completely limited to east Asia, and is entirely absent from western Asia and Europe – M175.

Today, M175, which arose on a Eurasian M9 chromosome, is found at highest frequency, around 30 per cent, in Korean populations. Based on absolute dating methods, it appears to be roughly 35,000 years old, coinciding very closely with the appearance of the Upper Palaeolithic in Korea and Japan. There are several more recently derived markers that have M175 as an ancestor (particularly M122, which will play a significant role in Chapter 8), and together these related lineages account for 60–90 per cent of the Y-chromosomes in east Asia today. Like a collection of soup recipes that all have a common ingredient, M175 unites most Asian men living east of the Hindu Kush and Himalayas, defining an east Asian clan.

When these modern humans reached east Asia, they found themselves in an area that had been inhabited by their distant relatives *Homo erectus* for nearly a million years. Dubois' missing link had relatives in China, called (before being united with their Javanese cousins to the south) Peking Man. But mysteriously, no *erectus* remains from Chinese sites are found after 100,000 years ago – there is a gap in the record until fully modern *Homo sapiens* make their appearance around 40,000 years ago. What caused this hominid gap is unclear, although the likely culprit is – once again – the steadily deteriorating climate. For example, the cave at Zhoukoudian, where many *erectus* remains have been found, is located in north-eastern China, near Beijing – a region that experiences extremely cold winters even today. During the intense cold of the penultimate glaciation, around 250,000 to 150,000 years ago, the climate in northern China would have become much harsher. Consistent with this, no *erectus* remains post-date 250,000 years at Zhoukoudian. It seems likely that the deep freeze drove them away – or even killed them off.

We know that *erectus* didn't change substantially for 1 million years in east Asia, perhaps the result of stable selection pressures. Isolation from other hominids and a penchant for relatively uniform climatic conditions would have favoured continuity rather than change, and there is no evidence for an *erectus* Great Leap Forward. While some Chinese scientists argue for an evolutionary model known as 'regional continuity', in which east Asian *erectus* evolved into a local variant of *Homo sapiens* independently of what was happening in Africa, there is absolutely no genetic evidence for this. Moreover, the genetic results show that there was not even any interbreeding between modern human immigrants to east Asia and *erectus* – if in fact any populations still existed 40,000 years ago that are invisible to today's archaeologists. In a recent analysis of over 12,000 men from throughout east

Asia, geneticist Li Jin and his colleagues found that every single one traces his ancestry to Africa within the past 50,000 years – because every man has our old friend M168 on his Y-chromosome. Everyone. This result is bad news to those looking for evidence of east Asian regional continuity, since it is impossible to reconcile with any form of local evolution from *erectus*, or even admixture – at least on the male line. East Asian mitochondrial DNA gives the same answer: the thousands of samples that have been tested all trace their ancestry back to Africa. In short, there is no genetic evidence that *Homo erectus* made any contribution to the gene pool of modern east Asians. Rather, Dubois' ape-man appears to have been an evolutionary dead-end, and he was completely replaced by modern humans.

If the story ended there, it would be very tidy and self-contained. But unfortunately, life is never that simple. In this case, the spanner in the works comes in the form of the presence of our Coastal lineage at high frequency in some east Asian populations. The Coastal lineage is found at a frequency of 50 per cent in Mongolia, and it is common throughout north-east Asia. How it reached this location remains a mystery, but it is likely that the early coastal migrants to south-east Asia gradually moved inland, migrating northward over thousands of years. The M130 chromosomes in the south are older than those in the north, consistent with such a migration. At some point, perhaps 35,000 years ago, they would have met the descendants of the other, main line of migrants – our incoming Eurasians. The presence of both Eurasian and Coastal lineages in east Asian populations attests to the extensive admixture that occurred between them.

The picture that emerges is that east Asia was settled by modern humans from both north and south, like migrational pincers or 'chopsticks'. The northern route, which was characterized by Eurasian clan members, probably entered around 35,000 years ago from the steppes of southern Siberia. The southern route, which was composed primarily of members of the Coastal clan, was probably in place before this – perhaps as early as 50,000 years ago. The present composition of east Asia still shows evidence of this ancient north–south divide. Luca Cavalli-Sforza, working with Chinese colleagues, examined several dozen non-Y-chromosome polymorphisms in east Asian populations. In their analysis, they saw a clear distinction between the northern and southern Chinese. Even members of the same ethnic group, such as northern and southern Han, are most closely related to their *geographic* rather than their *ethnic* neighbours; northern Han group with other, non-Han northern populations, and the southerners form a separate group. It seems that the ancient evidence of a two-pronged settlement is still visible in the blood of today's Chinese.

So, our Middle Eastern clan had made it to the eastern extreme of the continent. Along the way it had acquired additional markers, producing the widespread Eurasian clan, the Indian clan, and the central Asian clan. The mountain ranges of central Asia served as effective barriers to migration 40,000 years ago, as they continue to do today. The effect of this was to produce an isolated east Asian Y-chromosome clan that only occasionally pops up in the west. But while the route to eastern Asia was clear, that to Europe required a more circuitous tour. As we saw, modern Europeans contain rather too many ingredients in their soup to have been the direct descendants of the Middle Eastern clan. The search for the ancestors of the first Europeans is where we are headed next.

7
Blood from a Stone

Raven sent birds to pierce the wall of dawn; one of them pecked a hole, through which the rays of the sun shone for the first time. Then he scattered seal bones upon the earth in the morning light, and the bones assumed human form: the first man and woman.
Chukchi creation myth

When I was a postdoctoral research fellow at Stanford University, my future wife and I lived in San Francisco and commuted down 'the peninsula' to Palo Alto. We chose to make this trek every day because we preferred living in the city, with its excitement and fluid mix of people. Our flat was in the Richmond district, the heart of the Russian immigrant community and shoulder-to-shoulder with 'New Chinatown' on Clement Street. On the drive home at night I would listen to National Public Radio, which is more or less the American equivalent of BBC Radio Four, in order to pass the time. One evening in the autumn of 1997 as I was driving up 25th Avenue, I heard a news announcement that almost caused me to swerve into an oncoming bus. I pulled over and listened, hanging on every word.

The announcer was reporting that a team of scientists led by Professor Svante Pääbo of the University of Munich had just published the first DNA sequence from a Neanderthal. This was, in a way, one of the Holy Grails of anthropology – a research finding that promised to answer one of the oldest and most contentious questions in the field: had modern Europeans evolved from Neanderthal ancestors, or were the Neanderthals replaced by groups of humans invading from somewhere else?

Neanderthals were the first hominid ancestor to be discovered, at that cave in the Neander Valley back in 1856. In spite of the initial reluctance to accept the fact of human evolution, within a few decades most people came to believe that Neanderthals were the ancestors of modern Europeans. The genetic studies of the 1980s, however, called into question this theory. If mitochondrial DNA told us that everyone had come out of Africa relatively recently, how could modern Europeans have evolved from a hominid like the Neanderthal, present in Europe from around 250,000 years ago? It was a contentious question, with anthropologists such as Milford Wolpoff of the University of Michigan insisting that the DNA evidence was wrong, and that Europeans were Neanderthal under the skin.

The only problem with inferring details about the past from data collected in the present is, as we saw in Chapter 2, that you need to make use of theories of how DNA sequences actually change over time. These theories, although supported by generations' worth of genetic and evolutionary research, are still theories. Unfortunately, it's impossible to go back in time and check the evidence in order to confirm whether our theoretical inferences are valid. Or is it? Can we study the DNA of our long-dead ancestors?

The field of ancient DNA research was pioneered in the 1980s by Svante Pääbo and his colleagues (including Allan Wilson, of mitochondrial Eve fame) in Berkeley and Munich. The impetus behind this work was to do the impossible – to go back in time by examining the DNA that existed in a long-dead individual. It was, in effect, an attempt to develop a genetic time machine that would allow us to answer questions about our ancestors directly. One of the first applications was in the analysis of DNA from Egyptian mummies, but soon people were trying it on fossils that were millions of years old. Michael Crichton's novel *Jurassic Park* was based on the heady early days of the field, when it seemed that anything would be possible – even getting intact dinosaur DNA from bloodsucking insects embedded in amber!

While the claims for successful retrieval of DNA from sources that were tens of millions of years old eventually proved unfounded, usually resulting from minute amounts of contamination by modern DNA, it was sometimes possible to retrieve DNA from more recent samples, or those that had been preserved in ideal conditions for tens of thousands of years. The frozen bodies of mammoths and ancient alpine travellers yielded analysable DNA, as did the dried remains from mummies and other desert-dwellers. Even then, the analysis was almost always limited to mitochondrial DNA, present in huge numbers of copies in every cell – making it more likely that one copy would have survived the Russian roulette of molecular degradation over the centuries. It was still extremely difficult to do this sort of analysis, though, because in most cases the molecules had completely disintegrated after death. This meant that negative results were far more common than positives – but the stories revealed by the tiny fraction of cases where DNA could be successfully extracted made the effort worthwhile. It was with this in mind that Pääbo's group had developed reliable ways of evaluating and extracting DNA from ancient samples, and his laboratory represented the state of the art in the early 1990s – they were the undisputed experts in the field.

The scientific coup that led to my near-death experience in San Francisco actually began with the very first Neanderthal bones to be unearthed. Comprising the so-called type specimen – the one against which all the others were judged by palaeoanthropologists – these bones had sat in a museum in Bonn for nearly 140 years when the Munich group was approached to do the analysis on them. Pääbo jumped at the chance, and his graduate student Matthias Krings performed the DNA analysis as part of his PhD thesis work. In over a year of tedious trial-and-error work, Krings gradually managed to extract enough intact mitochondrial DNA to create a 105-base-pair sequence. What he saw when he pieced it together was extraordinary. Krings relates the first glimpse of the 40,000-year-old DNA:

I basically knew the sequence by heart . . . and I was certainly able to spot a substitution [DNA sequence change] when I saw one. After looking over the first sequence, I had something crawl down my spine. There were eight substitutions in a region which usually had – at most – three or four. I thought, 'This is a very funny sequence.'

After painstakingly reproducing the result from a separate bone fragment, and duplicating the experiment in a laboratory on a different continent (to be certain that a contaminant in the Munich laboratory was not producing an experimental artefact), he accepted the validity of the sequence. By repeating the procedure several times, he eventually managed to obtain 327 base pairs of mitochondrial DNA sequence from the remains – enough to generate a statistically significant estimate of its evolutionary divergence. The sequence was clearly not from modern human mtDNA, but it didn't belong to an ape either. Rather, it came from a hominid that last shared a common ancestor with modern humans around 500,000 years ago. This date was consistent with what was predicted by palaeoanthropologists who had studied the dispersal of so-called 'archaic humans' from Africa into Europe, and it proved that Neanderthal was not the direct ancestor of modern humans. Rather, Neanderthals represented a local population of archaic hominids who were later replaced by modern *Homo sapiens* – with no detectable admixture. Of the thousands of human mitochondrial sequences that have been obtained from people all over the world, not one is anywhere near as divergent as Krings' Neanderthal sequence. Neanderthals fall well outside the range of genetic variation found in the human species – and therefore they represent a *separate* species. This early result has been confirmed by two additional genetic studies of Neanderthal remains from different parts of Europe, showing that the Neanderthals were closely related to each other, but very distantly related to us. The genetic data is incontrovertible – modern Europeans trace their recent ancestry to Africa, in common with everyone else in the world.

Along with the study of 12,000 Asian Y-chromosomes discussed in the last chapter, the Neanderthal results placed the final nail in the coffin of multiregionalism. Our hominid relatives were clearly replaced by modern humans who spread out of Africa within the past 50,000 years. While there are still a few anthropologists who argue for a multiregional model of human evolution, most have accepted that there simply is no compelling evidence for it. The ghost of Carleton Coon has finally been laid to rest by modern molecular biology. But, you might be thinking, if the Neanderthals were replaced, who replaced them?

An artistic temperament

In the autumn of 1922 two teenaged boys entered a cave near Cabrerets, France (a two-hour drive north-east from Toulouse). Against the advice of their parish priest, with whom they had first entered the cave in 1920, they were intent on exploring it more fully. What they saw there was extraordinary. The paintings at Pech Merle (as the cave was christened) were later called the 'Sistine Chapel' of the region by Abbé Henri Breuil, the French expert on ancient cave art. His detailed research on dozens of French caves was to reveal a rich artistic tradition dating back over thirty millennia, giving a unique insight into the minds of Palaeolithic Europeans.

The images painted and drawn on the walls here and at other Upper Palaeolithic sites in Europe show clear evidence of conceptual, abstract thought – the earliest such evidence in the world. The extraordinarily detailed artwork at Chauvet cave has been dated to around 32,000 years ago, the oldest in France. Recently discovered drawings at Fumane cave, near Verona in northern Italy, may date from as early as 35,000 years ago, which would make them the oldest examples of cave art anywhere in the world. In all these locations, the complexity of the subjects, and the skill with which they are rendered, marks an abrupt transition from the past. In effect, they are detailed time capsules left by early Europeans – beautifully crafted snapshots of their lives, hidden away inside sealed caves until they were discovered in the nineteenth and twentieth centuries.

The inhabitants of these European caves were clearly talented artists, and their culture marks a distinct departure from that of the Neanderthals that preceded them. It marks the beginning of the Upper Palaeolithic in Europe, and broadcasts the arrival of fully modern humans on the scene. Along with the diverse tools they left, their art gives us a fleeting glimpse into the minds of the people who created it. But were these first European artists – the creators of Pech Merle, Chauvet and Fumane – the ancestors of western Europeans? And if so, why did they appear on the scene so suddenly, around 35,000 years ago? The genetic data gives us the clues we need to solve this puzzle.

We saw earlier that the most obvious location from which to enter

Europe, the Middle East, appears to have contributed little to the gene pool of modern Europeans. The Y-chromosome lineage defined solely by M89, which would have characterized the earliest Middle Eastern populations around 45,000 years ago, is simply not very common in western Europe. It is such a tiny hop across the Bosporus from the Middle East to Europe that we might ask why it took so long – perhaps 10,000 years – for modern humans to make a significant foray into western Europe. To solve this riddle of where the majority of Upper Palaeolithic Europeans came from – we need to examine the genetic markers in western Europe and ask which Eurasian lineage they could have come from, and when.

I said at the beginning of Chapter 5 that my Y-chromosome is defined by a marker known as M173. It turns out that this marker is not unique to me – in fact, it is found at high frequency throughout western Europe. Intriguingly, the highest frequencies are found in the far west, in Spain and Ireland, where M173 is present in over 90 per cent of men. It is, then, the dominant marker in western Europe, since most men belong to the lineage that it defines. The high frequency tells us two things. First, that the vast majority of western Europeans share a single male ancestor at some point in the past. And second, that something happened to cause the other lineages to be lost.

Desperate for a date

The first thing most of us want to know when we hear that almost all western Europeans trace their family line back to one man is 'when did he live?' This is where our absolute dating methods come in. If we examine genetic variation – polymorphisms – on the M173 chromosomes, we can estimate how long it would have taken for our mutational clock to create it. But if all of the chromosomes are M173, how can we study variation? Surely they are all identical?

Fortunately for us, they are not. While all of them are very closely related, and thus share the M173 marker, there are other markers that help to distinguish them. Unlike the stable markers we have studied to define the order – or relative dates – of the Y-chromosome lineages, these other markers do not involve simple one-letter changes in the genetic code. Rather, they exist because of a biochemical speech impediment. When we replicate our DNA, the double strands of the molecule open up and tiny machines known as polymerases actually do the hard work of assembling the complementary copy. Remember that if we know the sequence of one strand of the double-stranded DNA molecule, then we also know the other, because of the inviolable rules of molecular biology. A always pairs with T, and C always pairs with G. This works very well for more than 99 per cent of the genome, where the letters occur in a unique order and it is easy to tell how the pairing should work. Unfortunately, a small fraction of our genome is not so simple. It consists of what are known as tandem repeats – short sections of the same sequence, repeated several times in a row in the DNA strand. These often take the form of a couple of letters, such as CACACACACA . . ., but there can be three, four or more letters in the motif that is repeated. As you might expect, the polymerase can become confused when it encounters these parts of the genome. After all, if there are a dozen or more repeats, how can you tell where you are in the sequence – is it repeat number ten or eleven? So, in a reasonable number of cases (about one in every thousand), the polymerase makes a mistake when it is assembling the complementary strand, and adds or subtracts a repeat. If the original strand had twelve repeats, the copy may have eleven or thirteen – simply by chance, due to an error at the molecular level. It is a process that Luca Cavalli-Sforza has called genetic 'stuttering'.

One in a thousand may not seem like a very common event, but it is when we are talking about the work of DNA copying. If that was the rate at which polymerases made single-letter copying mistakes, then we would introduce an average of over a million mistakes, or mutations, into our DNA every time it was copied. Since genetic copying takes place when we are having offspring, this means that each child would be born with over a million new mutations. Biology takes a dim view of this level of mutation, and it is likely that the child would die of a horrible inherited disease – if it were born at all. Thus, the usual rate at which new mutations appear is much more sedate, perhaps twenty or thirty per generation. This is around 100,000 times lower than the mutation rate we see for repeats, which means that new mutations in 'regular' sequences are much less common than those in repeats. The repeats are on an evolutionary speedway, accumulating diversity at an extraordinary pace.

While this has very little effect on the health of the child, since repeats are usually found in regions of the genome that do not affect well-being, it does give us a tool for studying diversity. These repeats, known as microsatellites, are particularly valuable when we want to ask questions about variation on lineages where we do not have much single-letter variation – such as the M173 chromosomes. They give us a way to determine absolute dates that we can use to test our hypotheses about the timing of human migrations. The rate at which mutations occur is roughly constant, so the level of variation tells us how long they have been mutating. This tells us how old the chromosome is, because all of the chromosomes descend from a single chromosome at some point in the past. By definition, the level of variation at this point was zero, since there was only one copy.

When several microsatellites from M173 chromosomes are examined, the level of variation is consistent with an age of around 30,000 years. Of course, as with any estimate of time, this has a substantial range of error, but the most likely date for the origin of M173 is around 30,000 years ago. This date means that the man who gave rise to the vast majority of western Europeans lived around 30,000 years ago – consistent with a recent African diaspora, and again showing that Neanderthals could not have been direct ancestors of modern Europeans.

Significantly, it is around this time that the Upper Palaeolithic becomes firmly established in Europe – and the Neanderthals disappear. While the Chattelperronian interlude of around 38,000 years ago represented a short experiment with modernity, it is only after 35,000 years ago that we see the inexorable march of modern humans and their toolkits throughout the whole of Europe, as signalled by the appearance of the so-called Aurignacian stone tool industry. By 30,000 years ago the Neanderthals had been nearly eradicated, or perhaps reduced to isolated pockets such as those at Zafarraya in Spain. By 25,000 years ago they had disappeared entirely. The coincidence of the genetic and archaeological dates, as well as the increase in population size implied by the large number of Upper Palaeolithic sites from around 30,000 years ago, suggests that the invading moderns actually displaced the Neanderthals. But did we actively kill off our distant cousins as we spread into Europe?

Babies and grannies

A great many theories have attempted to explain the ultimate demise of the Neanderthals. Perhaps the most obvious, given the coincidence of the archaeological and genetic dates for the arrival of modern Europeans, is that they were killed by the newcomers in some sort of hominid genocide. There is, in fact, very little evidence for this. No prehistoric battle sites have been found in France or Spain, and there is little evidence of butchery on the Neanderthal skeletons that have been unearthed. Of course, archaeology may have missed the Neanderthal Waterloo, but on the face of it there is no evidence to suggest inter-species warfare. Rather, it was probably natural selection that did them in.

One of the things that the incoming Upper Palaeolithic Moderns had in their favour was a complex social structure. As we have seen, this probably began as an adaptation to cooperative hunting on the savannahs of east Africa. With their improved toolkits and bands of intelligent, social hunters, modern humans were much more efficient at hunting than the Neanderthals. This can be seen in the Neanderthal remains that have been found, all with extensive evidence of a harsh and physically difficult lifestyle. Most Neanderthals had broken bones, and many had quite extensive injuries that would have made them much less efficient members of the group. What modern humans accomplished with tools and brains, Neanderthals seem to have done with brute force. It was this physically demanding lifestyle that made them relatively short-lived. Few Neanderthals lived to be fifty, and most died in their thirties.

Neanderthals had always had a very dispersed social structure, with a small number of distinct groups, each with its own local tool-making variants. Some anthropologists have even suggested that different Neanderthal groups may have spoken different languages, which would have contributed to the fragmentation of their population. Whether this is true or not, the dispersed, nuclear nature of the

Neanderthal population probably represented an adaptation to the relatively harsh conditions of northern Europe during the last ice age. It allowed them to make use of the resources found over a wide territory, increasing their chances of locating food. It also, inadvertently, probably led to their demise.

Anthropologist Ezra Zubrow has calculated that a reduction in fertility, or an increase in mortality, of 1 per cent would have led to the extinction of the Neanderthals within 1,000 years. This degree of change is entirely consistent with a model in which Neanderthals were gradually excluded from their food resources by incoming, highly efficient Upper Palaeolithic humans. As they became squeezed into smaller and smaller territories, the Neanderthals would have been less likely to obtain the resources they needed in order to survive. Eventually, their numbers reduced through attrition, they may even have had difficulty finding mates. Admittedly this is all conjecture, but it is entirely consistent with the data on the time of arrival of the first Upper Palaeolithic Europeans, the mitochondrial DNA evidence for their population expansion beginning around 30,000 years ago, and the disappearance of the Neanderthals at the same time.

One feature of modern human behaviour that may have played a role in giving them an advantage over the Neanderthals was a by-product of the complex behavioural adaptations of modern humans. What probably began as Upper Palaeolithic hunting skills spilled over into complex social networks. This, coupled with their less physical lifestyles, would have given them a longevity advantage over the Neanderthals. Many Upper Palaeolithic people survived into their fifties, well past reproductive age. This gives us another clue as to why the Neanderthals were replaced: old people are good to have around.

A reliance on teaching and learning, rather than instinct, is one of the things that distinguishes humans from other animals. Most of our early lives are spent learning, and it isn't until we are well into our twenties that most of us feel that we are in command of sufficient knowledge to be able to synthesize and teach others. The older we get, the more knowledge we accumulate, and the more we can help our offspring to benefit from our experience. Grandparents, like university professors, have 'been there and done that' – and, crucially, lived to tell the tale. Having grandparents around also allowed higher fecundity, since (as any new parent can tell you) they can care for children while younger generations go about their lives. This includes continued childbearing – perhaps allowing that small advantage over Neanderthals that led to their extinction. Anthropologist Kristen Hawkes has suggested that grandmothering – the act of a child being cared for by its grandmother – may have played a substantial role in the population expansion of modern humans. Perhaps the small advantage it gave allowed modern humans to drive the Neanderthals to extinction.

Stepping-stones

Whatever the causes of their demise, Neanderthals had given up the ghost within a few thousand years of the arrival of modern humans. After 30,000 years ago, the only remains found in Europe are those of fully modern humans – often called Cro-Magnons, after the rock shelter in south-western France where some of the first bones were unearthed in 1868. These early Europeans were much more gracile, and significantly taller, than their Neanderthal neighbours. While Neanderthals were typically only around 165 cm (5 ft 6 in) tall, Cro-Magnons were often over 180 cm (6 feet), with long limbs. To palaeo-anthropologists such as Erik Trinkaus, these proportions suggest an origin in a much milder climate. Neanderthals, as long-time residents of the colder regions of Europe, had quite stocky and muscular proportions. The implication is that the Cro-Magnons arrived in Europe from somewhere warmer.

As we saw, lineages belonging to the Middle Eastern clan – which we would expect to find if there had been a straight shot out of Africa to Europe, via the Middle East – are hardly found at all in Europe. M173, our 30,000-year-old marker, has the advantage of being present at very high frequency in the most isolated European populations (including the Celts and the Basques), and its age corresponds roughly to the inferred date of modern human settlement based on archaeology. Other major Y lineages present in Europe are younger than M173, and thus arrived later, or descended from M173 itself. Thus M173 is the likely marker of the first modern Europeans, defining the European clan. Of course, it is simply the terminal marker in a long line of genealogical descent that traces back to M168 and our African Adam. The penultimate marker, though, actually solves the mystery of where the earliest Europeans came from. This marker, a stepping-stone on the way to M173, is M45 – making Europeans a subset of the central Asian clan.

As we discussed earlier, the steppelands of 30–40,000 years ago stretched across a vast swathe of the Eurasian landmass. To Upper Palaeolithic hunters, this ecosystem would have been a land of plenty, and migration along it would have allowed modern humans to disperse well to the west, into Europe proper, as well as to the east into Korea and China. During this period, the steppe zone extended well into present-day Germany, and may have reached France. We know from bones that have been found in French caves of 30,000 years ago that reindeer – a species adapted to the cold steppe and tundra of northern Eurasia – were common in France around this time. The climate had opened a window into Europe that allowed these central Asian steppe hunters to enter. As we have seen, they soon took over, dominating the region within a few thousand years.

It is likely that their sojourn on the steppes had honed their hunting skills, leading to innovations in technology that gave them a greater advantage over the Neanderthals than would have been possible if they had simply shot straight out of Africa. During the thousands of years they spent on the grasslands of central Asia they almost certainly underwent a period of intense cultural adaptation to this difficult environment. This period took the place of the hundreds of thousands of years of Neanderthal biological adaptation – what had given them their short, stocky frames. As recent migrants from tropical Africa, Upper Palaeolithic humans initially would have been ill equipped for life in the northern hemisphere. The central Asian steppes served as their apprenticeship, in a sense – preparing them for life in the most inhospitable environments on the planet. The caves of western Europe must have seemed relatively benign after the howling winds of the frozen Kazak grasslands.

It is this honing process that may explain why the early Middle Eastern immigrants did not come to dominate Europe. While the mountains and forests of the Balkans would have been a bit of a barrier for a species adapted to the steppes, some early Middle Eastern immigrants clearly did get through. We can speculate that the low frequency of their Y-chromosome lineages belies a population that was not quite ready for the rigours of life in western Europe – but of course it is impossible to say with certainty. What is clear is that most European men, including me, trace their ancestry back to central Asia within the past 35,000 years. And interestingly, this links us with a small population of Siberian hunters who – just as the last ice age was at its most intense – headed into the frozen tundra of north-eastern Asia.

The final frontier

Zaliv Kresta, the Bay of the Cross, is perched on the eastern edge of Russia, 10,000 km from Moscow. For six months of the year it is frozen solid, a mass of sea ice that isolates the small ex-Soviet settlement of Egvekinot from the rest of the world. The only way to reach it is by a two-hour helicopter flight from Anadyr, the nearest city with regular air connections to the outside world. From Egvekinot, it is a further eight-hour trek on military personnel carriers – armoured, with full tracks – inside the Arctic Circle to reach the reindeer herders living there. It feels like one of the most remote places on earth.

The people who live in this harsh environment, known as the Chukchi, are wonders of adaptation. They have developed a lifestyle that allows them to exist in an environment of unimaginable harshness. When I visited them in November 2001 the temperatures were already plummeting to −50°C at night, and in the depths of winter they can reach −70°. The landscape is an other-worldly tundra, covered in snow and frost from September to June, and there is no edible vegetation. The Chukchi live entirely off of their reindeer and the fish they catch through holes in the icy rivers. They manage to do this with technology that has changed very little over the past few thousand years, sewing their clothes from reindeer skin and sinew, living in tents constructed of hides and wooden poles and migrating with their herds as they search for the succulent lichen tips that provide their only source of nourishment.

Most of us who live in relative comfort in the modern world find it difficult to imagine how humans could exist in these conditions. And yet they do live – and thrive – in a climate that would probably kill most of us. Of all the hominids that have existed over the past few million years, it is only fully modern humans who have been able to live in the harsh Arctic. The conditions are simply too extreme to allow any mental leeway. Natural selection has favoured only those intellectually capable of surviving in this icy evolutionary laboratory.

This is certainly why we see evidence of human occupation in the Asian Arctic only after 20,000 years ago. If modern humans reached southern Siberia around 40,000 years ago, as the genetic and archaeological data suggest, it would take them another 20,000 years before they had developed the cultural adaptations necessary to live in the harsh conditions of the Arctic. It is also possible that population pressures, which may have encouraged a northward migration, were not felt until this time. Whatever the reason, the earliest north-eastern Siberian sites, such as that at Dyuktai, south-east of Yakutsk, and Ushki Lake, in Kamchatka, date from after 20,000 years ago. The people living in Siberia during this time appear to have developed a tool-making culture that was distinct from that of populations living further to the south and west, consistent with their highly adapted lifestyle. They were particularly adept at making microliths, small weapon points, by striking both sides into a symmetrical 'leaf' shape. Similar types of stone points have also been found in the earliest excavated American sites, suggesting a direct continuity in culture between Siberia and the Americas.

Anthropologists had assumed for many years that Native Americans and Asians have a common origin. Thomas Jefferson even stated the case in his 1787 book *Notes on the State of Virginia*:

. . . if the two continents of Asia and America be separated at all, it is by only a narrow strait . . . and the resemblance between the Indians of America and the Eastern inhabitants of Asia, would induce us to conjecture, that the former are the descendants of the latter, or the latter of the former . . .

Several anthropological traits – most famously the dental pattern known as sinodonty – are found in north-east Asia and the Americas. By the mid-twentieth century anthropologists such as Carleton Coon had even begun to classify Native Americans as 'Mongoloid' in their racial checklists. The problem was that no one knew exactly how long the Native Americans had been living there, and when they had split from their Asian cousins. In the 1950s carbon dating was used to infer an age of 11,000 years for the archaeological site at Clovis, New Mexico. The remains at Clovis contained leaf-shaped stone spear points in the same layer as extinct mammoth bones, which immediately suggested great antiquity to its discoverers. Over the next two decades, sites dating to roughly the same time period were excavated throughout North America. The pattern that seemed to be emerging from the archaeological record was that humans had occupied the Americas no earlier than 12,000 years ago.

In the 1970s and 1980s, though, three archaeological digs – one in North America and two in South America – turned up evidence for a human presence before Clovis. The Meadowcroft Rockshelter in Pennsylvania yielded artefacts that were originally dated using radio-carbon to roughly 14,000 years ago, pre-dating Clovis by 3,000 years. The care with which Meadowcroft was excavated was impressive, and while the dates for the earliest occupation have been revised downward (to around 12,500 years ago), they are accepted by many anthropologists. The site at Monte Verde in southern Chile yielded similar dates to those at Meadowcroft, roughly 13,000 years ago, although nearby hearths have been estimated to be as old as 33,000 years. The earlier date has not been widely accepted, and thus Monte Verde is thought to date – like Meadowcroft – to around 13,000 years ago.

The age of the remains at Monte Verde suggests that humans must have been in North America at least several hundred years prior to reaching Chile, and so pushes back the date of settlement a bit. But it was the final site that was the real bombshell. In a 1986 paper in the scientific journal *Nature*, archaeologist Niede Guidon summed up the find in the title: 'Carbon-14 dates point to man in the Americas 32,000 years ago'. It was the result of her excavation of the cave of Boqueirão de Pedro Furada in north-eastern Brazil, and it seemed to pull the rug out from under the post-13,000 consensus. Careful examination, though, has failed to confirm Guidon's results. The charcoal from the site, which provided a radiocarbon date and was thought by Guidon to be the remains of a fireplace, could have been produced by a natural fire. Furthermore, most of the crude stone artefacts discovered there do not look convincingly human in origin – they could easily have resulted from natural breakage. These doubts have led palaeoanthropologist Richard Klein to suggest that 'Furada may soon join the long list of dubious claims [for early human settlement in the Americas]'.

In summary, most of the reliable archaeological evidence points to a settlement of the Americas within the past 15,000 years. There is one small problem with this scenario, however: it was at this point that the ice age was at its most intense, and if early humans came from Siberia – as the anthropological and archaeological material suggests – they would have had to traverse the harshest environment on earth just as it reached its nadir. It implies a journey of unimaginable hardship for a species that only recently left its tropical homeland. Surely this would have been impossible? It is at this point that the genetic data provides us with more clues.

Doug Wallace, a geneticist at Emory University in Atlanta, had helped to pioneer mitochondrial DNA analysis of human populations when he was at Stanford University in the early 1980s. By the time he moved to Emory in the mid-1980s, he had become focused on the origins of Native Americans. In particular, he was trying to use mtDNA as a tool to track the origin of the first Native Americans back to particular populations in Asia. The first major publication of this work, in 1992 with Antonio Torroni, showed that Native Americans could be divided into at least two waves of migration. The earliest led to the settlement of both North and South America, while the later wave of migration left genetic traces only in North America. Their estimates of when these migrations took place varied widely, and could have occurred any time between 6,000 and 34,000 years ago. The results confirmed, though, that Native Americans and north-east Asians shared a recent common mitochondrial ancestry.

But how do these results fit with our Y-chromosome data? This question was answered in 1996 by Peter Underhill and his colleagues. Underhill found a single nucleotide change on the Y-chromosome, later named M3, that was common throughout the Americas. While their sample of Native Americans was by no means exhaustive, over 90 per cent of South and Central Americans they examined were M3, while around 50 per cent of North Americans had this lineage. Clearly, it was the major Native American Y-chromosome founder, defining the American clan.

The only problem was that M3 was not found in Asia. This could have been due to its age, which Underhill and his colleagues estimated to be as little as 2,000 years. The age estimate was extremely uncertain, however, as it was made in the early days of Y-chromosome analysis, and the mutation rate of the single microsatellite used to assess M3 diversity (using the same method that we used to date M173 in Europe) was uncertain. Thus it could also have been as much as 30,000 years old. Clearly more work was needed.

This came in 1999, when Fabricio Santos and Chris Tyler-Smith at Oxford and Tanya Karafet and Mike Hammer at the University of Arizona independently reported that the ancestor of M3 was defined by a marker called 92R7, named for an undefined nucleotide change on the Y-chromosome. They found that 92R7 was present in populations throughout Eurasia, with a distribution from Europe to India. In combination with other nucleotide changes this pinpointed Siberia as the source population for Native Americans, confirming Wallace's results from mitochondrial DNA. It was difficult to assess the age of the 92R7 lineage, however, as it was so widespread. What was needed was an additional marker on the lineage that would focus attention on the precise populations that could have given rise to the first Native Americans.

When it was later shown that M45 marks the same Y lineage as 92R7, the results made much more sense. Here was our central Asian marker, the same one that gave rise to M173 in Europe. It seems that the central Asian clan had made it to the New World as well, picking up the defining M3 marker in the process. This helped to trace a clear migrational route from Africa to the Middle East to the Americas, via the Eurasian steppes, but it still left us with the problem of how to date the first entry into the Americas. It could have happened any time between 40,000 and 12,000 years ago, taking into account both the genetic and archaeological results.

A recent analysis of the M45 lineage by Mark Seielstad and myself has defined a further marker, known as M242, which is a descendant of M45. It appears to have arisen in central Asia or southern Siberia around 20,000 years ago, and is distributed across Asia, from southern India to China to Siberia, as well as throughout the Americas. It is found at highest frequency in Siberia, and thus it could be called a Siberian marker. It is also immediately ancestral to M3, and defines an evolutionary order of M45 → M242 → M3 that traces a migration from central Asia to the Americas within the past 20,000 years. M242 appears to be the oldest genetic marker in the Americas. Thus the Y-chromosome results have given us much the same picture of the founding of the Americas as the mtDNA results, but have narrowed down the date of entry considerably. Clearly an entry prior to 20,000 years ago is inconsistent with the genetic results, since M242 was still in central Asia at that time. A more recent migration from Siberia is overwhelmingly likely, consistent with the archaeological evidence.

The picture that seems to be emerging from the genetic analysis of Native Americans is that of a migration by the Siberian clan from southern to eastern Siberia within the past 20,000 years. This initial move established a population at the north-eastern edge of Asia. Adapted to a hunting life on the central Asian steppes, they would have subsisted almost entirely off of the large mammals of the far north – musk ox, reindeer and mammoth among them. Consummate hunters, with finely crafted microlith tools, portable dwellings and clothing capable of withstanding the intense cold, these well-adapted tundra dwellers would have gradually extended their range eastward. As the ice age moved toward its lowest temperatures, and more moisture became tied up in the ice caps, sea levels would have dropped by over 100 metres. This would have created a land bridge in Beringia, between Siberia and Alaska, of ice-free land formerly submerged in the Bering Sea. The Siberian clan would have been able to move back and forth across this connection, living a dual Asian–American existence.

However, these first Americans of 15–20,000 years ago had one more obstacle to overcome. They would almost certainly have been barred from southward expansion by a continuous sheet of ice that covered most of northern Canada and eastern Alaska. It was only as the ice age began to abate, after 15,000 years ago, that it would have

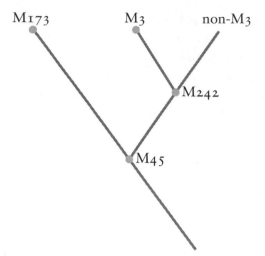

Figure 8 M45 is the ancestor of most western Europeans (who have M173) and Native Americans (who have M242 and M3).

become possible to transit the formerly icy interior and enter the North American plains, perhaps via a so-called 'ice-free corridor' that some palaeoclimatologists believe ran along the eastern edge of the Rocky mountains. It is around this time that grizzly bears first enter North America from Siberia, showing that humans weren't the only species to have been stopped by the Alaskan ice. So, the genetic age of 20,000 years, as well as climatological considerations having to do with the extent of glaciation and sea levels, provide an explanation for why we don't see archaeological remains in the Americas before this time. While archaeologists may someday discover a site that pre-dates 15,000 years ago, the mass of evidence is now in favour of a relatively late initial entry to the Americas. The stones and bones seem to agree with the DNA.

Manifest destiny

Interestingly, the Native American genetic data allows us to estimate how many people would have made it into the continent during these first migrations. By looking at the number of chromosomes needed to account for the present distribution of genetic lineages in the Americas, and working out how much diversity would have accumulated over the time that they have been in the continents, it is possible to account for all of the mtDNA and Y-chromosome types in Native Americans with a founding population of around ten or twenty individuals. Because some lineages would have gone extinct during the past 15,000 years, as we saw with our French soup recipes, this is certainly an underestimate of the number of individuals who actually made it across. Perhaps a few dozen, or even a few hundred, actually made the trip. Clearly, though, the diversity present in the Americas is a tiny fraction of that found in Eurasia – which in turn is merely a subset of that found in our African forebears. And of those who made it to Alaska, only a few left descendants. The gene pool of Native Americans carries in it a signal of the hardships faced by their Beringian ancestors thousands of years ago as they moved ever deeper into the deep freeze, scraping a living from the frozen wastes of the far north.

After they made it through the rigours of life in the freezer, the plains of North America must have seened like the promised land. Here was a vast grassland – much like the steppe they left long ago in central Asia – filled with large grazing animals. It was as though someone who had been adrift on a raft in the ocean for weeks was suddenly transported into a supermarket. The result was a massive increase in population as these highly efficient Siberian hunters took advantage of their newfound good fortune. In just 1,000 years or so they jour-neyed all the way to the tip of South America, in the process helping to kill off many of the species that made the plains such good hunting grounds. Three-quarters of the large mammals in the Americas were driven to extinction around this time, among them mammoths and horses – the latter weren't to reappear in the Americas until the Spaniards introduced them in the fifteenth century. While humans may not have done the job on their own – climate change at the end of the last ice age almost certainly played a major role – they probably delivered the *coup de grâce* to the gentle giants of the plains.

Counting waves

One of the most contentious issues in the study of Native American origins is deceptively simple: how many waves of migration were there into the New World? If the earliest Americans came from Siberia, did later migrants arrive from further afield? The 9,500-year-old 'Caucasoid' skull recently discovered at Kennewick, in Washington state, hints at ancient connections to Europe. Some anthropologists believe that Australian Aborigines migrated to South America, while others think that the Japanese managed to sail across the Pacific thousands of years ago. Can the genetic data help us to sort through these possibilities and weed out the plausible from the simply barmy?

Linguistics provides us with one clue. The languages spoken in the Americas – over 600 by some estimates – have long been a contentious issue for linguists. Are they related to each other, or is their diversity simply too great to be subsumed into a few language 'families'? American linguist Joseph Greenberg, who will play a role in the next chapter, suggested in the 1950s that the vast majority of the languages spoken in the Americas belong to a single language family, which he called Amerind. While this hypothesis has certainly not won universal acceptance, Greenberg has argued his case persuasively, and many scholars are beginning to accept it. Apart from Amerind, which includes all of the languages spoken in South America and most of those spoken in North America, linguists recognize two other language families: Eskimo-Aleut and Na-Dene. Eskimo-Aleut is spoken only in Greenland and the northern parts of Canada, as well as in Alaska and eastern Siberia, while Na-Dene languages are spoken in western Canada and the south-western United States. Do the language families give us a clue about the history of migration to the Americas?

Greenberg suggested that each family originated with a single migration from Asia to the New World. The speakers of each then spread the languages through the Americas as they migrated, producing the distributions we see today. This model implies that there should be some genetic correlation with the linguistic groups – after all, if it was the movement of people, rather than languages, that caused their spread, then genes should have moved as well. Recent genetic studies have provided support for Greenberg's classification, suggesting that there were indeed at least two waves of migration originating from different parts of Asia.

Greenberg thought that the Amerind family was introduced by the earliest migration into the Americas because it is the most widespread, and is the only one spoken in South America. The genetic data bears this out, with Amerind speakers in both North and South America sharing high frequencies of M242 and M3 – marking them as members of the Siberian clan. The mtDNA data obtained by Torroni and Wallace also supports an early Amerind settlement of the Americas. It seems likely that our Beringian hunters were speaking a language that was ancestral to modern Amerind languages, and that 12,000 years of divergence has produced the extraordinary linguistic variety we see today.

Since Na-Dene was the next most widespread family, Greenberg suggested that it was brought by a second wave of migrants. We do, in fact, see a genetic signal of this later migration. It comes, interestingly, in the form of our Coastal marker, M130. In Na-Dene populations, as many as 25 per cent of men have this marker, while it is found at much lower frequencies in neighbouring northern Amerind speakers. Tellingly, M130 is not found in South America. The genetic dates indicate that it migrated to the Americas within the past 10,000 years, originating in the region of northern China or south-eastern Siberia. By this time the Bering land bridge had been engulfed by the sea once again, so these migrants almost certainly came by boat, migrating along the coast. This is supported by the present distribution of the Na-Dene languages, which are limited to the western half of North America. It seems likely that their ancestors followed the coast all the way around the Pacific Rim, travelling as far as California. The distribution of the Na-Dene languages we see today reflects the continuation of a coastal migration that began in Africa around 50,000 years ago, moving eastward via India to south-east Asia and Australia before heading north towards the Arctic and the Americas. The Coastal marker reveals the deep relationships among the inhabitants of these far-flung places.

And what about the Eskimo-Aleut speakers? There does not seem to be a distinct genetic signature for this group, and it is likely that it arose as a subset of the M242-bearing Siberian clan, who took on a coastal lifestyle. They migrated to the east, as far as Greenland, using their kayaks to hunt walrus and seal – but their genetic lineages tie them back to their ancestors in Siberia, the tundra-dwelling mammoth hunters of 20,000 years ago.

As for the other migrations, from Europe or Australia, there is currently no compelling genetic evidence. While M130 would appear to link Na-Dene speaking Native Americans to the Australian Aborigines, the relationship is in fact far deeper, and reflects a common ancestry tens of thousands of years ago in south-east Asia. Likewise for Europeans, who share a common ancestor with most Native Americans, revealed by the high frequency of the central Asian M45 marker in both groups. Furthermore, since Siberians and Upper Palaeolithic Europeans initially came from the same central Asian population, they probably started out looking very similar to each other. Kennewick Man, as a likely descendant of the first migration from Siberia to the New World, may have retained his central Asian features – which could be interpreted as 'Caucasoid'. In fact, many early American skulls look more European than those of today's Native Americans, suggesting that their appearance has changed over time. The more 'Mongoloid', or east Asian, appearance of modern Native Americans may have originated in the second wave of migration, carrying M130 from east Asia. There is no evidence, however, for an M175-bearing migration of Chinese or Japanese sailors across the Pacific – this marker is simply not found in today's Native American populations. The genetic evidence is quite clear: all ancient migrants to the Americas seem to have travelled via Siberia.

Bang

By 10,000 years ago all of the world's continents (apart from Antarctica) had been colonized by humans. In just 40,000 years our species had travelled from eastern Africa to Tierra del Fuego, braving deserts, towering mountains and the frozen wastelands of the far north. Their ingenuity had stood them in good stead during this journey, and they had become exquisitely well adapted to life in conditions that were a far cry from their African birthplace. But just as these Upper Palaeolithic wanderers were settling into their new homes, something significant happened. Although it started out as a trivial experiment, it was to change for ever the way that humans interacted with their world. It could be called the second 'Big Bang' of human evolution and, like the Great Leap Forward, it would launch another human journey – this one into the realm of recorded history.

8
The Importance of Culture

When the world was first created and the gods were born, each deity had a task in the maintenance of the land. This hard labour led to complaints and demands to find a better solution. One day the water goddess Nammu decided to create man out of clay; Enki and Ninmah were given the task. They drank too much beer and began to play a game where one created beings and the other found a role for them. Three had malformed genitals, and became priests. One was completely unviable, unable to stand or feed itself, and had to be held in Ninmah's lap – the first human infant.
Sumerian creation myth

The Hawaiian archipelago lies in the middle of the Pacific Ocean, over 3,200 km (2,000 miles) from the nearest continental landmass, North America. Today it is one of the major tourist destinations in the United States, with millions visiting its beaches every year. The short flight from California, high-rise hotels and Honolulu traffic belie the isolation of the islands. Today native Hawaiians are a tiny minority in their homeland, but this is a phenomenon of the past hundred years – at one time they were one of the most isolated human populations in the world. And, like the Australians, it is clear that they must have arrived in Hawaii from somewhere else, since there are no other primate species living on the islands. The notion that they voyaged here by boat seems almost unthinkable, yet – like our Oz-bound coastal migrants of 50–60,000 years ago – they must have made the trip.

When Captain Cook arrived on the island of Kauai in 1778, he was unaware of the ancient voyage the Hawaiians had taken to arrive at this remote spot. He was leading a four-year expedition aboard the *Resolution*, exploring the north Pacific in an effort to discover the elusive north-west passage between the Atlantic and Pacific Oceans. Cook named the archipelago the Sandwich Islands, after his benefactor the Earl of Sandwich. The native Hawaiians, although interesting as anthropological specimens, were not accepted as equals – and their own name for their native land was ignored.

Cook noted the primitive character of the people living in Hawaii – in particular, the fact that they were still living in the 'Stone Age' and had neither the benefit of metallurgy nor written language. In fact, when he first encountered them, their incredulous reaction to the *Resolution*'s nautical equipment led him to infer that they had never been aboard a ship. Yet in spite of their apparently primitive way of life, the Hawaiians had made an epic sea journey in order to reach their home. And it was not unique: the nearest inhabited Hawaiian neighbours are the Marquesas Islands, 3,500 km to the south-east, and beyond that there is another 1,500 km of open ocean before reaching the Society Islands, still in the middle of the Pacific Ocean. If Hawaii had been settled by the most direct route of island-hopping, minimizing the distance travelled between each inhabited island, then there would have been at least two enormous sea passages in addition to many other shorter hops. Clearly this was no accident. The Polynesian seafarers who colonized Hawaii were accomplished sailors, able to travel between distant outposts of dry land throughout the Pacific without the benefit of compasses or clocks to infer longitude.

It is now generally accepted, based on the earliest archaeological evidence for a human presence in Polynesia, that these consummate seafarers made all of their voyages within the past 4,000 years. What led them to make the leap into the unknown world of the Pacific? And if humans had been capable of crossing open oceans since at least the time of the first Australians, why did it take them so long to colonize Polynesia? To find the answers to these questions we will have to take a trip back to Eurasia, in search of the factors that led up to the Polynesian odyssey.

A break with the past

The Tell el Sultan is 25 km north-east of Jerusalem, on the eastern slope of the Mountains of Judah. The Arabic word *tell* refers to a mound left by human occupation, and archaeologists have been digging there since the 1870s. Most were looking for evidence to support stories from the Bible, and the uppermost layers in Tell el Sultan do, in fact, belong to the biblical city of Jericho – the name most often used for the site. These later remains, dating from the past 4,000 years, were most carefully scrutinized, but during the course of their work the archaeologists uncovered evidence for earlier occupation. It was only with the focused work of Dame Kathleen Kenyon in the mid-1950s, though, that the earliest layers were systematically explored. What she found there would change our concept of human history.

Kenyon found evidence for human settlement at Jericho dating from around 10,000 years BC – hunter-gatherer communities that lived off of the game and water resources in much the same way as their Upper Palaeolithic ancestors had 30,000 years before. Then, immediately above this, she found the remains of an early farming community, dating from the period immediately afterwards. The plaster- and shell-decorated skulls she unearthed, evidence of an ancestor-worshipping cult, are some of the best-known artefacts in archaeology. These and other evocative remains made Kenyon one of the most famous archaeologists of her era, but it was the age of the settlements that were to have the greatest effect on the study of prehistory. Up to that time the first known villages had been dated to the fifth millennium BC, while true towns only started to appear 2,000 years later. Using radiocarbon methods, the lowest urban layers at Jericho were dated to around 8500 BC, meaning that this single excavation pushed back the date of the first permanent human settlements by 4,000 years. Kenyon's excavation of Jericho revealed the earliest evidence in the world for a settled, agrarian society.

In the modern world, with its densely populated settlements and reliance on farmed crops and domesticated animals, it is easy to forget that only a few hundred generations ago we were all hunter-gatherers. For most of us, life has changed so completely since the Palaeolithic that we imagine we have always lived as we do now. In fact, as the deep excavation trenches at Jericho show first-hand, there was a sudden transition from hunting and gathering to settled life around 10,000 years ago. What is particularly fascinating about the timing of this event is that it appears to have happened nearly simultaneously in several independent locations around the world. This suggests that there was a common reason for Upper Palaeolithic people to abandon their nomadic ways and settle into domesticated bliss.

The Middle Eastern culture that immediately preceded the earliest settled, or Neolithic, layers at Jericho belongs to a short-lived cultural tradition known as the Natufian, named after the first site where it was uncovered, Wadi an-Natuf in Israel. The Natufian economy centred on gathering cereal plants – particularly the ancestors of wheat and barley, which were plentiful in the Middle East around this time. It was the end of the last ice age, and the eastern Mediterranean was warming up. The improving climate encouraged the growth of large stands of cereals and nut-bearing trees at higher latitudes than during the ice age, allowing the Natufians to exploit these new resources. By specializing on plentiful species, they were able to settle in one place (near their favoured plants) and still gather enough food to survive.

Middle Eastern archaeologists have found that the end of the last ice age was a period of intense climatic variation in the eastern Mediterranean, with a general pattern of change from a continental to a Mediterranean climate. As archaeologist Brian Fagan summarizes it, this had the effect of producing an ecological zone with long, dry summers and short, wet winters. The effect of this climatic change was to favour grasses, which produce seeds in the spring and then lie dormant over the summer. Early humans would have exploited the relative plenty of food during the spring by harvesting large quantities of seed, then storing it for the rest of the year. This concentrated gathering behaviour would have favoured a settled lifestyle, which set the stage for the revolution that was to follow.

After 9000 BC the eastern Mediterranean summer was becoming increasingly drier as the full effects of rising global temperatures kicked in. This reduced the yield of cereals, and (as with arid periods in the distant past) would have favoured mobility. However, the necessity of storing their gathered grain would have tied the Natufians to one location. The pull of these two forces – reduced yields and a relatively immobile lifestyle – would, within a few hundred years, lead some Natufian settlements such as Jericho to try a new innovation: planting some of the gathered cereals (which are really seeds) in order to simplify the gathering process. Kenyon's work at Jericho traced the development of the Neolithic, or New Stone Age, following this early innovation. Archaeologists and anthropologists continue to debate what happened after the first crops were cultivated – whether the need for a reliable water supply in order to grow crops led to the development of irrigation, which may have fostered water-rights issues and social hierarchies, and so on. What is clear is that the end of the ice age appears to have set in motion a series of events that were to culminate with the development of settled, agrarian communities within a thousand years. What archaeologist V. Gordon Childe called the 'Neolithic Revolution' had arrived.

The second Big Bang

The Neolithic marked a turning point for the human species. It was at this point that we stopped being entirely controlled by climate – as we were during our Palaeolithic wanderings – and began to assume control of our own destinies. By adopting agriculture, Neolithic humans initiated several developments that characterize modern civilization. The first is that of choice. The Natufians living at Jericho made a conscious decision not to wander for miles each day in order to gather their food. Rather, they decided to mould their environment to suit themselves, modifying the natural state of nature in order to favour human behaviour. While some hunter-gatherers practised forms of environmental control (the Australians, for instance, burned scrubland periodically in order to favour the grassland animals they hunted), the early agriculturalists of the Middle East, China and America were directly controlling the species in question. This gave them more choices as to where they could live, and would allow them to thrive in areas that had proved marginal for hunting and gathering.

The second development was that of greatly increased population density. One of the consequences of cultivating food and settling in one place is that the necessity of not over-exploiting limited resources is relaxed – after all, if you want to have more children, you can simply plant and harvest more crops. While this does oversimplify the situation, it is true that settled, agrarian societies are more densely populated than those of hunter-gatherers. Coupled with the freedom to choose where to live, this can lead to very rapid population expansions, with agriculturalists spreading throughout a region. It is estimated by palaeodemographers, who study past population sizes using archaeological and anthropological methods, that the entire population of the globe was around 10 million at the time agriculture originated; by the dawn of the Industrial Age, around 1750, world population had risen to over 500 million. If Palaeolithic hunter-gatherer populations had taken over 50,000 years to increase from a few thousand individuals living in sub-Saharan Africa to a few million scattered around the globe, clearly the agriculturalists of the last ten millennia were making up for lost time.

The final new feature of the Neolithic revolution is that it demonstrates the importance of new technology to human migration. In much the same way as the central Asian steppe dwellers of 20,000 years ago used their technological superiority to occupy areas of Siberia that had been strictly off-limits to our hominid ancestors, so too did our more recent ancestors gain an adaptive advantage from technology. The first major technological development of the last 10,000 years was agriculture, and it would set in motion a massive acceleration of human social evolution. In fact, as we will see, it would be over 9,000 years before a similarly important series of developments would initiate another era of human evolutionary history. Clearly the development of agriculture was a pivotal event. If the Great Leap Forward had set the stage for the first Big Bang in human history, which led us to colonize the world, agriculture was to set in motion the second – one that would send our species hurtling into the modern age.

The genetic fallout

While agriculture played a critical role in the development of modern society, the genetic effects of agriculture were equally pronounced. While Upper Palaeolithic hunter-gatherers tended to maintain a relatively stable population size, except via the settlement of new territory, agrarian societies were able to expand massively without leaving home. As the first farming communities increased in size, their inhabitants gradually moved further afield in search of cultivable land. When they did so, they carried with them their genetic markers. One of the consequences of this is that we see the expansion of certain genetic lineages, giving us a glimpse of the origin and spread of agriculture. In the case of the Middle East, the genomes of today's western Eurasians still retain a signal of those events at Jericho 10,000 years ago.

Archaeologists had long known that agriculture spread from its origin in the Middle Eastern 'Fertile Crescent' to Europe over the course of several thousand years. The earliest evidence is found in the Balkans, and it appears later and later as you move to the north-west. It is only relatively recently that ancient Britons left behind their hunter-gatherer lifestyle, several thousand years after their cousins in Jericho had done the same. Crucially, it is exactly those plant species initially cultivated in the Fertile Crescent that make their appearance in the advancing wave of agriculture as it moved into Europe. It seems that the European hunter-gatherer lifestyle was replaced by the new Middle Eastern invention.

In the 1970s Luca Cavalli-Sforza, along with fellow geneticists Alberto Piazza and Paolo Menozzi, initiated a study of the genetic effects of agriculture. The question that they asked was about the way in which agriculture had spread. In particular, they wanted to know if the migration of agriculture into Europe marked the migration of people, or simply the spread of a sexy new cultural development – the MTV of its era. In effect, they were asking a question about the genetic composition of modern Europeans. Was there evidence for an expansion of certain genetic markers out of the Middle East, or did modern Europeans appear to be relatively free of Neolithic markers?

At the time the study was done, the only data available was that on the 'classical' markers we learned about in Chapter 2 – blood groups and other cell-surface protein markers that served as convenient polymorphisms but gave little information about their underlying DNA sequence changes. The analysis of these markers led Cavalli-Sforza and his colleagues to conclude that there had been a mass migration of genes out of the Middle East, and the genetic pattern was very similar to that observed for the first appearance of agriculture: the genetic signal decreased regularly as you moved from south-east to north-west across Europe. The methods of analysis used in this study limited what the researchers were able to infer, since it wasn't possible to obtain an accurate date for this migration, but their findings did corroborate the theory that agriculture had spread with farmers as their population had expanded, rather than as a purely cultural phenomenon that 'migrated' as Palaeolithic Europeans learned farming skills.

Cavalli-Sforza's results became accepted wisdom, leading to what they called the 'Wave of Advance' model for the diffusion of agriculture. The assumption made by many (although not Cavalli-Sforza and his colleagues) was that the majority of the European gene pool was Neolithic in origin, since it was the most pronounced genetic pattern in Europe (although Cavalli-Sforza's later work showed that it still accounted for less than a third of the genetic variation). Many anthropologists remained sceptical, but it was to be over twenty years before the model received a serious re-evaluation. This came in the late 1990s with the detailed analysis of mtDNA lineages in Europe and south-west Asia by Martin Richards and his colleagues at Oxford University. In a series of scientific papers they analysed mtDNA lineages from a selection of populations across Europe and the Middle East, carefully dating the lineages using the absolute methods similar to those we learned about earlier. This allowed them to evaluate the relative contributions of various migrations to the European gene pool. Their results suggested that, rather than having a significant genetic effect on the population of Europe, the expansion of agriculture involved very few Middle Eastern migrants. Most of the lineages in Europeans seem to have been present since the time of the Upper Palaeolithic, between 20,000 and 40,000 years ago.

One of the objections to the Richards study, raised by Cavalli-Sforza and others, was that mtDNA actually provided very little resolution between European populations. It was difficult, for instance, to distinguish between eastern and western Europeans with mtDNA data alone – they have very similar patterns of mtDNA markers. Nonetheless, the mtDNA result was suggestive. What was needed was to look at the male lineage, with its greater inherent resolution, in order to see if it showed the same pattern.

This was finally done in 2000, when Ornella Semino and her colleagues (among them Cavalli-Sforza) analysed the Y-chromosomes of over 1,000 men from Europe and the Middle East, looking specifically for evidence of the agricultural expansion. What they found was that lineages defined by Neolithic Middle Eastern markers are found in a minority of modern Europeans. In fact, the results from the Y agree almost perfectly with the mtDNA data, suggesting that 80 per cent of the European gene pool traces back to other waves of migration, primarily during the Palaeolithic. In western Europeans, this Palaeolithic component is largely defined by our friend from the last chapter M173, which links Europe back to central Asia. Only 20 per cent of European Y-chromosomes -- defined by more recent markers, particularly one known as M172 – derive from Neolithic Middle Eastern immigrants. In effect, modern Europeans are largely genetically Cro-Magnon on both their maternal and paternal sides.

This is not to say that the advent of agriculture had no effect on Europe – far from it. There is clear genetic evidence for a significant European population expansion after the end of the last ice age, almost certainly aided by the onset of food production. Evidence for this comes in the form of a recent analysis by David Reich and his colleagues at Massachusetts Institute of Technology. They studied markers from many independent regions of the genome and found a pattern of variation suggesting that the European population underwent a substantial decrease in population size between 30,000 and 15,000 years ago, as Europe was moving into the depths of the last ice age. This was then followed by a population expansion from the few survivors after the end of the last ice, producing the relative dearth of variation seen in Europe today. In other words, the human population had been through what is known as a bottleneck – a reduction in size followed by a period of growth. Patterns of mtDNA variation also support this model of postglacial population growth. Archaeological evidence suggests that the Palaeolithic population of Europe was confined to Iberia, southern Italy and the Balkans during the period of most extensive glaciation around 16,000 years ago, and that human populations then expanded northward during the postglacial period. Agriculture almost certainly contributed to the end of this population expansion, because it allowed much higher population densities.

How do we reconcile the pattern seen for the Y-chromosome and mtDNA, of a Palaeolithic European population relatively unaffected by Neolithic immigration, with the Wave of Advance? The pattern seen by Cavalli-Sforza and his colleagues clearly exists, but they were studying large-scale patterns across the entirety of Europe and the Middle East. The agricultural expansion was simply one population movement into Europe – there is clear archaeological evidence for several others. As their later analysis showed, it still accounted for a minority of the genetic variation in Europe. Furthermore, because the Wave of Advance had no estimate of age, the Neolithic component could have been confounded with Palaeolithic immigration from the Middle East. Finally, since central Asian populations were not included in their analysis (there was no data available at the time their study was conducted), it is possible that the pattern reflects a general trend of migration from Asia to Europe during the Upper Palaeolithic. After all, if we simply had Y-chromosome data from the Middle East and Europe, we would infer that M89-bearing populations had migrated into Europe via the Balkans, giving rise to M173 in Europe. It is only because we know that M173 arose on an M45-containing lineage that we trace the Upper Palaeolithic settlement of Europe back to central Asia.

The Y data actually provides a partial solution to the apparent conundrum. It seems that southern European populations experienced a greater influx of Neolithic farmers from the Middle East, carrying lineages such as M172, than did northern Europeans. One possible scenario is that farming spread first around the Mediterranean, with Neolithic Middle Eastern immigrants favouring its climate, similar to that of the Levant. Only later did indigenous Palaeolithic Europeans take up agriculture in the interior, gradually spreading the culture – but only a small percentage of the genes – of the Neolithic throughout. The Cro-Magnons of northern Europe appear to have made a conscious decision to leave behind the Palaeolithic for a new Middle Eastern lifestyle with a small minority of Middle Eastern immigrants.

Rice Man

While the complexity of Neolithic spread in Europe makes a simple interpretation of the genetic data difficult, the situation in the other major centre of Asian domestication is a bit clearer. The pattern of settlement and intense exploitation of a few plant species that characterized the Middle East was seen at around the same time in China. Northern Chinese sites such as Banpo and Zhangzhai in Shaanxi province show early evidence of millet agriculture, around 7000 BC. Millet, a cereal crop like wheat, seems to have been domesticated around the Yellow River, spreading from there to the rest of northern China. The remains at Pengtoushan, on the Yangtze River in central China, indicate that rice was being cultivated there independently around the same time. At both sites pottery was used for storing grain, and the people lived in carefully constructed clay houses, suggesting that the Neolithic lifestyle was well developed, even at this early date. Agriculture soon spread throughout China, with rice dominating in the south, where the wet, humid conditions favoured this grain. Rice agriculture spread down the Yangtze, and was widespread in southern China by 5000 BC, perhaps helped by a second, independent domestication along the south coast. By 3500 BC it was being cultivated in Taiwan, and by 2000 BC in Borneo and Sumatra. It reached the rest of the Indonesian archipelago by 1500 BC. Overall, the archaeological evidence suggests that rice agriculture spread from an origin in central-southern China to the islands of south-east Asia in the space of roughly 3,000 years – timing similar to that of the agricultural expansion into Europe. However, unlike in Europe, there is a very strong genetic signal of this expansion, suggesting that it was people, and not merely the culture, that moved.

In Chapter 6 we learned that one descendant lineage of M9, defined by a marker known as M175, is widespread in east Asia. Based on its present distribution, this marker probably arose initially in northern China or Korea. Looking at the pattern of Y variation in modern

Chinese populations, it is now clear that the first agriculturalists in China were descendants of M175. In fact over half of the entire male population of China have Y-chromosomes defined by a marker that shows evidence of a massive expansion in the past 10,000 years. M122, which first appeared on an M175 chromosome, is now widespread throughout east Asia. It is hardly found west of the great central Asian mountain ranges, and does not occur at all in the Middle East or Europe. This is the pattern we expect to see with a recent expansion, rather than an ancient event that typically leaves a more widespread trail.

The genetic data shows that the development of rice agriculture in east Asia created a Wave of Advance. However, while the wave leaving the Fertile Crescent for Europe seems largely to have dissipated after inundating the Mediterranean, the one leaving China was to saturate the entirety of east Asia. Today, M122 – marking the descendants of the first Chinese rice agriculturalists – is found from Japan to Tahiti. A recent study by David Goldstein and his colleagues at University College London shows that microsatellite diversity on M122 chromosomes is very high in China and Taiwan, but drops significantly moving southward into peninsular Malaysia and Indonesia. This is precisely what would be expected from a population expansion originating in China within the past 10,000 years – and exactly parallels the archaeological evidence for the spread of rice agriculture. Together, M122 and a related Chinese haplotype (also a descendant of M175) defined by marker M119, account for nearly half of the Y-chromosomes in south-east Asia. In Europe, on the other hand, Neolithic immigrants account for only 20 per cent of the present Y diversity. In comparison to Europe, the Wave of Advance in east Asia appears to have been more of a tsunami.

Double-edged scythe

The massive population expansion made possible by the adoption of agriculture, whether it was through population growth of the originators themselves (as in east Asia) or the people who adopted agriculture from them (as in much of Europe), suggests that this innovation was nothing but good news. After all, if success is broadcast through excess, a massive increase in agricultural populations must have meant that life improved after the Neolithic transition. Recent evidence suggests that this may not have been the case.

Early agriculturalists were taking on a new set of risks when they committed themselves to a settled existence. The most important was a decrease in the breadth of their resource base. By focusing cultivation on a few species, they were reducing their choices in the event of a climatic shift. Droughts, intense periods of cooling (such as the Dryas periods at the end of the last ice age) and shifts in watercourses were all very easy to deal with for Palaeolithic hunter-gatherers. Their response to any of these changes was to move into another area with better resources. The great migrations of Palaeolithic humans – those covered in Chapters 4 to 7 – were almost entirely determined by the climate. Once humans adopted agriculture, though, they were loath to move. This led to occasional famines, such as those seen today in many parts of the developing world. In the early days of agriculture, during the turbulent climatic conditions of the early postglacial period, famine episodes would have been even more likely.

The second main worry for our Neolithic agriculturalists was the increase in disease. While hunter-gatherers may appear to have had a difficult life, relying as they did on apparently 'primitive' technology and the necessity of killing or gathering enough food to survive, in fact they were surprisingly healthy. While the incidence of broken bones and wounds is greater for Palaeolithic humans than for their sedentary Neolithic descendants, they do not appear to have died younger. In fact, the skeletal remains from early agricultural communities suggest that early agriculturalists may actually have had a *shorter* lifespan than their hunter-gatherer neighbours. This is thought to be due largely to an increase in disease.

Infectious diseases do not arise spontaneously as a by-product of a settled lifestyle, but rather from exposure to disease-causing organisms in such a way that transmission occurs from one infected individual to another. Most diseases can exist only in large populations, where a threshold number of people remain infected, allowing the disease to remain in the population. These are so-called endemic diseases, such as smallpox or typhoid. A population of several hundred thousand is necessary to maintain the disease – otherwise it is lost because not enough people remain susceptible to infection. Populations of this size only arose after the development of agriculture. Other diseases can be introduced from an outside source, such as an animal. While humans had contact with animals as hunter-gatherers, the sort of prolonged, close contact that encourages the spread of disease occurred only after the domestication of animals in the Neolithic. Measles, for instance, is closely related to rinderpest, a disease of cattle. It is likely that the domestication of livestock around 10,000 years ago introduced this disease into Neolithic populations. Historian William McNeill has suggested that many of the plagues described in the Bible may have had their origin in the outbreaks of epidemic disease during the early days of the agricultural transition in Eurasia.

The final negative aspect of a sedentary lifestyle was the growing stratification of society. In general, hunter-gatherers are remarkably egalitarian, having few social divisions. Typically, taking modern-day populations such as that of the San or Australian Aborigines as a model, there is a group leader who sits in judgement over some aspects of group life, but no formalized set of social divisions such as the ones that exist in settled societies. Perhaps because there is simply less to fight for (in terms of accumulated wealth), large-scale warfare is rare in hunter-gatherer societies – although inter-group battles do occur. The massive growth in population during the Neolithic created conditions in which some form of social stratification was inevitable. Once this occurred, the seizure of power and the growth of empires was not far behind, which led to war on a scale that had never been seen in the Palaeolithic. And while warfare was bad enough on its own, it also had a knock-on effect on other aspects of Neolithic life. The high mortality associated with large-scale warfare was probably exacerbated by the spread of disease and the destruction of cropland during the hostilities, leading to a vicious chain reaction of mortality.

Given all of the negative aspects of the Neolithic revolution, why did our ancestors still embrace their new lifestyle? Not everyone did, in fact – small pockets of hunter-gatherers existed in almost every region of the world until quite recently. Their reasons for maintaining an ancient lifestyle probably had something to do with the environment (for instance the San and the Australian Aborigines live in marginal, arid environments that are difficult for agriculture), as well as a conscious decision to remain hunter-gatherers. For the rest of the world's population, though, there was no turning back. It is possible that the shift in thinking that allowed humans to accept agriculture, in spite of all its negative aspects, could have occurred in a few generations. Once the collective memory of hunting and gathering was replaced by one involving food production, it would have been virtually unthinkable to return to the old ways. Ask yourself if you would be prepared to make weapons and hunt for your dinner – most of us would probably say no.

Babbling

The onset of the Neolithic established many of the regional patterns of cultural diversity we see in the modern world. Expanding waves of agricultural migrants in east Asia spread rice cultivation to Indonesia and beyond, and today their descendants still carry the genetic traces of this event. As we saw earlier, the first inhabitants of south-east Asia may have been more similar to today's Andamanese or Semang Negritos. It is likely that most of these groups were engulfed by the wave of expanding rice-growers, their culture subsumed into the agricultural mainstream. Similarly, hunter-gatherer groups in Europe, the Americas and Africa all gave up their Palaeolithic lifestyle in favour of the new way of feeding themselves. But culture is defined by far more than eating – it encompasses social traditions, clothing and tool-making styles, means of transport and thousands of other things. And one the most important aspects of culture is language.

Most American visitors to Britain soon notice the huge number of regional accents. If London is the first stop, then the Cockney accent will be one of the first ones they encounter. Even if they've been practising their Dick van Dyke equivalent ('Cor blimey, Mary Poppins!'), it is sometimes difficult for them to believe that the same language is being spoken. My English wife finds it similarly perplexing to talk to some of my American friends from the South. George Bernard Shaw was right when he noted that the Americans and British are two people separated by a common language – and he wasn't even taking into account local variation within each country. Accents are familiar examples of language variation, and the difficulty we may have in understanding them reveals an insight into the process of language change. Languages are not uniform entities, in spite of the efforts of the Académie Française to impose order on the rowdy French populace. As with any aspect of culture, there is a great deal of variation from place to place. But does the apparent chaos of linguistic diversity reveal anything about the spread of human cultures?

Language similarities had been recognized since Classical times, particularly among such well-studied European examples as Latin, French, Spanish and Greek. By the eighteenth century scholars had begun to take a broader view, focusing on the languages of Asia, Africa and the Americas. For instance Janos Sajnovics, in his obscure 1770 treatise 'Demonstration that the language of the Hungarians and Lapps is the same', arrived at the conclusion given in the title. We now know that both Hungarian and Lapp belong to something known as the Uralic language family, uniting them with more obscure languages such as Khanty, Nenets and Nganasan. Sajnovics, though, was unaware of these more distant relationships. And while he, like many other scholars, recognized the similarities uniting different languages, he crucially failed to explain *how* they had arisen.

An explanation for the similarities among members of a language family arrived a few years after the Sajnovics study. In a 1786 address to the Royal Asiatic Society, Sir William Jones – then a judge in India – noted that Sanskrit (the religious language of Hinduism) bore a closer resemblance to Greek and Latin 'both in the roots of verbs and in the forms of grammar, than could possibly have been produced by accident'. So much so, he concluded, that they must 'have sprung from a common source'. It was this last statement that was to be his most lasting contribution, since it implied a mechanism for the generation of linguistic diversity. Languages change over time, he was saying, and if there are enough deep similarities among a group of languages, then they must have had a common ancestor in the past and subsequently diverged from each other. It was an evolutionary explanation for linguistic diversity, presaging Darwin by over sixty years.

The languages that Jones described all belong to what became known as the Indo-European language family, after the geographic locations of the languages. There are 140 separate languages in the family, ranging from those belonging to the Celtic branch, spoken in the extreme north-western parts of Europe (Gaelic and Breton are two examples), to Sinhalese, spoken in Sri Lanka. English is a member of the Germanic branch of Indo-European, although its complicated history has left it with many words borrowed from French. Clearly, this is a widespread and diverse collection of languages.

Today the hypothesis that Jones advanced – that all of the Indo-European languages trace their descent from a common ancestor – is widely accepted by linguists. In fact it is one of the few language families to have received universal acceptance. The implication of his model, known as the *genetic* model of language classification, is that at some point in the past there was a group of people who spoke an ancestral form of Indo-European, which later evolved into the languages we see today. Like our soup recipes, additions and modifications of ingredients have produced local linguistic varieties, which eventually became distinct languages. The parallels with DNA evolution seem obvious. But is it possible to learn anything about language diversity – and to understand the present distribution of the world's languages – from the study of genetics?

The subject of language change has always been a key interest of Luca Cavalli-Sforza, particularly its overlap with genetic patterns. Instead of drawing vague comparisons between genetic and linguistic diversity, in 1988 he decided to test the hypothesis directly – much like Dick Lewontin had done with the genetic data from different races. He and his colleagues examined genetic data from forty-two worldwide populations and drew a tree of their relationships based on minimizing the differences in marker frequencies between them. The tree that resulted – in effect, a genealogical tree of the populations – corresponded very well with known linguistic relationships. So, for instance, speakers of Indo-European languages tended to group together in the genetic tree, as did speakers of Bantu languages in Africa. There were obvious inconsistencies, such as the deep split between northern and southern Chinese (almost certainly resulting from the pattern of early migrations discussed in Chapter 6), but overall the genetic and linguistic groups seemed to be very similar to each other. This suggested that genetic data could be used to study the origin and dispersal of languages.

There were two caveats made by Cavalli-Sforza and his colleagues in their study. The first is that the genetic markers they were studying did not *cause* the pattern of linguistic diversity – there was no Bantu gene that forced its hapless carriers to speak those languages. Rather, similar genetic markers reflected the common history of the speakers of that language, as markers of descent. The second caveat is that in many cases relationships suggested by genes and languages disagreed, showing that the correspondence wasn't absolute. The reasons for this might be language replacement, in which people learn to speak a new language without a corresponding influx of outside genes, or gene replacement, in which there was a significant influx of genes but the language stayed the same. The first explains the difference between northern and southern Han Chinese, while the second may explain the close genetic similarity between linguistically unrelated groups, such as Na-Dene-speaking Native Americans and neighbouring Amerind speakers. Thus, genes were often markers of linguistic relationships, except when they weren't. Either way, the genetic data should help to shed light on language relationships, by illuminating the way in which languages have spread.

In search of a homeland

If we accept that William Jones was right, and that all Indo-European languages descend from a common source, then the implication is that there must have been, at some point in the past, a single group of people who spoke this ancestral form of Indo-European. The search for the identity of the first Indo-Europeans, and their geographic location, has been one of the main areas of archaeological and linguistic investigation over the past 200 years. It has become a sort of quest – although like all good quests, it is in some ways rather quixotic. The attempt to disentangle the web of conflicting evidence surrounding the location of the Indo-European 'homeland' illustrates a particularly exciting new application of genetics to our understanding of human history.

Gordon Childe, who coined the term 'Neolithic revolution', proposed in the 1920s that the Indo-European homeland should be identified with a culture originating north of the Black Sea that had distinctive 'corded' pottery – marks that resembled impressions left by cord or twine. The theory was revived by archaeologist Marija Gimbutas in a series of articles published in the 1970s. Gimbutas argued that the remains left by nomadic horsemen of the southern Russian steppes, dating from around 6,000 years ago, mark the earliest signs of a culture that can be identified as proto-Indo-European (PIE), which included Childe's Corded Ware people. The Kurgan culture, as she called it, left enormous burial mounds (known as kurgans) that are still dotted across the entirety of the Eurasian steppe, from Ukraine to Mongolia and south to Afghanistan. The golden treasure hoards recovered from kurgan excavations in the twentieth century confirmed the existence of a people who were known to Herodotus as the Scythians – fearsome horsemen of the Asian grasslands, and previously thought by many scholars to be mythical.

The evidence that the Kurgan people spoke PIE is based on an analysis of words common throughout Indo-European languages. If a word can be shown to derive from the same root, then it is likely (although not certain) that it was inherited from the common ancestor. For instance, the English word *ox* has cognates in the Sanskrit *uksan* and the Tocharian (an early Indo-European language spoken in western China) *okso*. Similarly, many words for animals and plants are common throughout the Indo-European languages, as are those for tools and weapons. Perhaps most interestingly, there is a rich vocabulary for horses and wheeled vehicles in common among all the languages, suggesting that the PIE speakers had domesticated the horse as a draft animal. Coupled with the archaeological remains showing that the horse was domesticated in the southern Russian steppes, this pointed toward the kurgan-builders being the PIE people.

But while the evidence in favour of the Kurgan people being early Indo-Europeans was compelling, there was no archaeological evidence for the spread of their culture into western Europe. Their culture, dominated by horses, was ideal for the steppes, but it was not well suited to European forests and mountains. It was difficult to see why the steppe horsemen would have been able to conquer Europe and impose their language upon its inhabitants. For this reason, Colin Renfrew proposed in his 1987 book *Archaeology and Language* that the Kurgan culture did not mark the origins of Indo-European, but rather a later, eastern extension of it. Renfrew suggested that PIE had been a Middle Eastern language, originally spoken 9,000 years ago, which spread with the agricultural Wave of Advance into Europe. He identified Anatolia as the Indo-European homeland, on the basis of it being roughly central in the modern distribution of Indo-European languages, and also the home of several extinct examples. The hypothesis he advanced is that the early farmers carried their language – PIE – with them as they expanded their population, and thus the linguistic inundation of Europe should have involved a genetic wave as well. It was a bold suggestion, which had little initial support from the linguistics community. As we have seen, the Wave of Advance actually contributed little to the gene pool of modern Europeans, and its influence seems to have been largely limited to the Mediterranean region. The Indo-European speakers living in Ireland, for instance, have virtually no Neolithic Y-chromosome markers, while Greeks have a substantial Neolithic component. What this suggests is that, if farming spread Indo-European languages throughout Europe, it must have done so largely without the actual spread of farmers – thereby reducing the strength of Renfrew's argument.

Of course, as the name suggests, Indo-European languages are spoken not only in Europe. Modern Iran, Afghanistan and the Indian subcontinent all have a majority of Indo-European speakers. How did they come to speak languages related to Irish Gaelic, thousands of miles away? Again, there are competing hypotheses. The first, advanced by Childe, Gimbutas and others, is that the early steppe horsemen carried their language from central Asia into India when they invaded around 1500 BC. The Rig Veda, an early Indian religious text, records the conquest of India by mounted warriors from the north. This received corroboration in the 1920s when Sir John Marshall and his colleagues excavated Mohenjo Daro and Harappa in the Indus Valley. These great cities date from around 3500 BC, and by the second millennium BC they were massive settlements with thousands of houses, extensive agriculture and enormous populations. Then, around 1500 BC they entered a period of decline, and by AD 1000 the Harappan culture had disbanded, its cities abandoned. What caused this sudden cultural collapse? To the archaeologists, it seemed to correlate perfectly with an invading force of Aryans from the Steppes. Archaeology seemed to be reinforcing Childe's argument, and corroborating the Rig Veda.

More recent research has suggested that there were probably in-digenous causes for the collapse of the Harappan civilization. Perhaps a river changed course, or social decay had set in (think of the Romans, 2,000 years later). Whatever the cause(s), the invading Aryans were not necessarily the all-powerful conquerors that early archaeologists thought they were. In the wake of this reinterpretation, Renfrew sug-gested two models for how the Indo-European languages could have come to India.

Renfrew's first model is that of an early Neolithic migration from the Middle East, with the settlers carrying their PIE language with them. In this model, the Harappans would already have been Indo-European, and thus there is no reason to infer an Aryan invasion in order to account for the languages of India. The second model, giving more credence to the Rig Veda, is that there was an invasion of the Indus region by Indo-European speaking nomads from central Asia, but it was carried out by relatively few individuals. Thus it had little impact on the population of the subcontinent, aside from the impo-sition of a language and culture. In both cases, the Indian genetic data shows a minor contribution from the northern steppes.

The test of the Childe–Gimbutas and Renfrew hypotheses awaited the development of markers that were capable of distinguishing between populations from the steppe and the indigenous Indian gene pool. As we saw in Chapter 6, M20 defines the first major wave of migration into India from the Middle East, around 30,000 years ago. It is found at highest frequency in the populations of the south, who speak Dravidian languages – a language family completely unrelated to Indo-European. In some southern populations, M20 reaches a frequency of over 50 per cent, while it is found only sporadically outside India. Thus, for our purposes, it is an indigenous Indian marker. What was needed to complete the analysis was a steppe marker, in order to see what contribution it may have made to the genetic diversity present in India.

This came with the discovery of a marker known as M17, which is present at high frequency (40 per cent plus) from the Czech Republic across to the Altai Mountains in Siberia and south throughout central Asia. Absolute dating methods suggest that this marker is 10–15,000 years old, and the microsatellite diversity is greatest in southern Russia and Ukraine, suggesting that it arose there. M17 is a descendant of M173, which is consistent with a European origin. The origin, distribution and age of M17 strongly suggest that it was spread by the Kurgan people in their expansion across the Eurasian steppe. The key to solving our language puzzle is to see what it looks like in India and the Middle East.

The answer is that M17 in India is found at high frequency in those groups speaking Indo-European languages. In the Hindi-speaking population of Delhi, for example, around 35 per cent of men have this marker. Indo-European-speaking groups from the south also show similarly high frequencies, while the neighbouring Dravidian speakers show much lower frequencies – 10 per cent or less. This strongly suggests that M17 is an Indo-European marker, and shows that there was a massive genetic influx into India from the steppes within the past 10,000 years. Taken with the archaeological data, we can say that the old hypothesis of an invasion of people – not merely their language – from the steppe appears to be true.

And what of the Middle East? Interestingly, M17 is not found at high frequency there – it is present in only 5–10 per cent of Middle Eastern men. This is true even for the population of Iran, speaking Farsi, a major Indo-European language. Those living in the western part of the country have low frequencies of M17, while those living further east have frequencies more like those seen in India. What lies between the two regions is, as we learned in Chapter 6, an inhospitable tract of desert. The results suggest that the great Iranian deserts were barriers to the movement of Indo-Europeans in much the same way that they had been to late Upper Palaeolithic migration.

The Y-chromosome results from Iran and the Middle East also suggest that early Middle Eastern agriculturalists did not spread Indo-European languages eastward as they moved into the Indus Valley. The marker M172, associated with the spread of agriculture, is found throughout India – consistent with an early introduction from the Middle East, most likely during the Neolithic. But the frequency is comparable in Indo-European and Dravidian speakers, suggesting that the introduction of agriculture pre-dated that of the Indo-European languages. Thinking in terms of actual behaviour, many Indian descendants of Neolithic farmers have learned to speak Indo-European languages, while fewer M17-carrying Indo-European speakers – up to this point – have given up their language in favour of Dravidian.

The low frequency of M17 in western Iran suggests that, in this case, exactly the sort of scenario envisaged by Renfrew in his second model has occurred. It is likely that a few invading Indo-European speakers were able to impose their language on an indigenous Iranian population by a process Renfrew calls *elite dominance*. In this model, something – be it military power, economic might, or perhaps organizational ability – allowed the Indo-Europeans of the steppes to achieve cultural hegemony over the ancient, settled civilizations of western Iran. One candidate for this 'something' was their use of horses in warfare, either to pull chariots or as mounts. Cavalry and chariots, both steppe inventions, would have given the early nomadic Indo-Europeans a distinct advantage over their adversaries' infantry. The use of horses would provide a major technological advantage to armies over the next three millennia. It is not difficult to imagine that it gave an early advantage to the people of the Eurasian steppe.

Thus, while we see substantial genetic and archaeological evidence for an Indo-European migration originating in the southern Russian steppes, there is little evidence for a similarly massive Indo-European migration from the Middle East to Europe. One possibility is that, as a much earlier migration (8,000 years old, as opposed to 4,000), the genetic signals carried by Indo-European-speaking farmers may simply have dispersed over the years. There is clearly *some* genetic evidence for migration from the Middle East, as Cavalli-Sforza and his colleagues showed, but the signal is not strong enough for us to trace the distribution of Neolithic lineages throughout the entirety of Indo-European-speaking Europe. Cavalli-Sforza has suggested that an initial migration of Neolithic pre-PIE speakers from the Middle East could have introduced a language to Europe, including our Kurgan people, which later became PIE. There is nothing to contradict this model, although the genetic patterns do not provide clear support either.

There is another possibility, which comes from the distribution and relationships among extinct languages in the Middle East and Europe. What if the language of the first farmers was not Indo-European, but another language entirely? The Basques, who live in north-eastern Spain, speak a language unrelated to any other in the world. Jared Diamond, in his book *The Rise and Fall of the Third Chimpanzee*, suggested that it might be a remnant of the agricultural Wave of Advance from the Middle East. Interestingly, some linguists have suggested that Basque is related to languages spoken in the Caucasus, while others find similarities to Burushaski, a language isolate spoken in a remote part of Pakistan. Similarly, there were other now-extinct languages spoken throughout the Mediterranean world, in south-eastern Spain (Tartessian and Iberian), Italy (Etruscan and Lemnian) and Sardinia (there is a non-Indo-European source for many place names). Place names in southern France similarly suggest that Basque was much more widely spoken in the past than it is today, and Greek place names indicate the presence of a pre-Indo-European element there as well. Overall, there is reasonable evidence for a 'Mediterranean' collection of pre-Indo-European languages that were later replaced by the expansion of Greek and Latin.

Taken at face value, then, we have a set of languages that were once widespread around the Mediterranean and Middle East, extending eastward into Pakistan. This is precisely the territory colonized by early Neolithic farmers during the period between 10,000 and 7,000 years ago. One possibility is that these early farmers spread 'Mediterranean' languages as they expanded their populations. The Palaeolithic populations of Europe took on the language of farming, and its culture, even if (as in the case of the Basques) there was hardly any genetic influx. These languages also spread to the east, introducing farming throughout the river valleys of central Asia and Pakistan. Later migrations, of Dravidian and Indo-European speakers in the case of Pakistan, and Indo-Europeans in the case of Europe, would have reduced the current speakers of the Mediterranean languages to the isolated pockets we see today.

Of course, this scenario is purely speculative, but it may be a plausible alternative to Renfrew's Indo-European farmers and Cavalli-Sforza's pre-PIE farmers. Furthermore, the genetic data shows some correlations: most of the regions mentioned, from the Mediterranean to the Caucasus to Pakistan, have substantial frequencies of M172, our canonical Neolithic marker. This is particularly true for populations from the Caucasus, some of which have frequencies of M172 in excess of 90 per cent. The generally close genetic similarly between Caucasian populations and those from the Middle East suggests that there was a substantial influx of people during the Neolithic, who may have introduced languages related to Sumerian to the region. Of course, this scenario assumes a relationship among all of the Mediterranean languages, which is tenuous at best. However, some linguists have found evidence for such a language 'superfamily', revealing deep structures common to seemingly unrelated languages. The search for these superfamilies is where we are headed next.

The big picture

Charles Darwin, writing in the time before modern methods of language classification had been fully worked out, noted the similarity between classifications based on genealogy and those based on linguistics. In the *Origin of Species*, he noted that 'if we possessed a perfect pedigree of mankind, a genealogical arrangement of the races of man would afford the best classification of the various languages now spoken throughout the world'. Cavalli-Sforza has said that he was unaware of Darwin's hypothesis when he began his 1988 comparison between genetic and linguistic relationships, his attention having been drawn to it later by a colleague who studied the history of science. It is not, perhaps, such a great leap of faith to suggest that languages tend to track population relationships. After all, we do 'inherit' our language from our parents, so at least on the time scale of the recent past, languages should be a good proxy for genes. What happens, though, when we look further? Is there a deeper relationship among languages that unites them into larger groups? And, perhaps most importantly, is there any evidence of a linguistic equivalent of our genetic Adam or Eve?

Joseph Greenberg, whom we encountered in Chapter 7, was convinced that such deeper relationships did exist. He made his name in the field of linguistic classification by uniting the hundreds of languages of Africa into four distinct families, described in his 1963 book *The Languages of Africa*. These early attempts at higher-order classification were generally well received by the linguistics community, and their success encouraged Greenberg to begin to look tentatively at deeper relationships among languages, particularly those of Eurasia.

Greenberg found that many of the languages, including those belonging to the Indo-European family, seemed to share certain structural elements that they found too striking to be due to chance. The details seem trivial to non-specialists (one example is how nouns are made into plural forms, by the addition of either a -*k* or -*t* suffix), but are significant to many linguists. Merritt Ruhlen, in his book *The Origin of Language*, traces many of the similarities among Greenberg's so-called Eurasiatic family, called Nostratic by some specialists.

One of the first questions we might have about this group of languages is whether, like Indo-European, there is any archaeological or genetic evidence for it. Unfortunately, this does not appear to be the case. One problem is that its members are so widespread across much of Eurasia that it encompasses a huge number of distinct populations. This may be due to the estimated age of the family – perhaps more than 20,000 years old. Any correlation with such an ancient and widespread group of languages is tenuous at best, and the only obvious Y-chromosome marker would be M9. M9, however, is also found in the other Eurasian superfamily of languages, known as Dene-Caucasian.

The first group in this family is that of the American Na-Dene languages (such as Navajo) and Sino-Tibetan, the languages of China and Tibet. Many linguists now accept the relationship between these two language families, but the more distant relationships are much more controversial. This is because Dene-Caucasian also includes, as its name suggests, languages from the Caucasus, as well as Basque and Burushaski. To put this into perspective, the languages belonging to Dene-Caucasian are spoken from the Pyrenees to the Rockies, with isolated patches scattered across Eurasia – a rather disparate group to say the least. In part because of this, American linguist John Bengtson has identified a subgroup within Dene-Caucasian that includes Basque, Caucasian, Burushaski and the extinct Sumerian language. The overlap with our hypothetical 'Mediterranean' family is striking, and (as we have already seen) there is some genetic evidence to support the dispersal of this group of languages during the past 10,000 years, perhaps in association with agriculture. The inclusion of Sumerian is especially telling, since this language – spoken by one of the earliest Mesopotamian civilizations – has geographic and cultural links back to the earliest days of agriculture in the Fertile Crescent.

While the genetic data supports the notion of a population connection among some of the western members of the Dene-Caucasian family, there is no clear link between them and the eastern languages. These languages, the Sino-Tibetan and Na-Dene families, do have their own genetic connection, however. It comes in the form of the M130 marker, which we first encountered in tracing the coastal migration to Australia. As we saw in the last chapter, M130 is also found in the population of eastern Asia, including China, marking a northward expansion of the marker from south-east Asia. Interestingly, this marker is also found in Na-Dene-speaking populations in North America. As with the Na-Dene languages themselves, it is not found in South America. This suggests a unique genetic link between east Asians and some Native American tribes, which arose from a second migration into the Americas between 5,000 and 10,000 years ago. In this case, genetics reinforces the linguistic relationship and provides a rough date for the divergence.

Their success in identifying common features in languages separated by tens of thousands of years has led some linguists to delve even further into the recesses of linguistic history, searching for the deepest relationship of all – a common origin for all languages. Merritt Ruhlen, one of the staunchest supporters of this view, believes that the Dene-Caucasian family marks the earliest spread of modern humans out of Africa, while the Eurasiatic family marks a later expansion emanating from the Middle East. As we have seen, there is no clear genetic data to support this model. One alternative is that these families spread, at least in part, via cultural dissemination, without leaving well-defined genetic trails. This has happened with some branches of Indo-European, for instance. The other possibility is that Eurasiatic and Dene-Caucasian do not really exist – perhaps they are simply collections of unrelated languages that show random similarities. Or perhaps subgroups do exist, particularly those supported by genetic data (such as Sino-Tibetan and Na-Dene), while many of the languages are unrelated. Ruhlen clearly has his work cut out for him.

It is likely that the evolution of language does follow the same paths as the migration of modern humans, with an origin in Africa and subsequent dispersal to the far corners of the globe. However, this statement is based on circumstantial evidence – the universality of language in all human populations, extrapolation from short-term linguistic change in recognized families such as Indo-European, and the presumed importance of language for the development of modern human culture. Almost all of the signals of the original human language – if it existed – have been lost, leaving us with today's dispersed Tower of Babel. In the same way that English fragmented into a large number of dialects that became more dissimilar over the past 500 years, so too do all languages become more dissimilar over time. Eventually, they lose all evidence of their common origin. The period of time required for this is unclear. Some linguists think that 6,000 years is long enough, while Ruhlen and others claim to have found similarities that trace back more than 20,000 years. The search for the language of Adam and Eve promises to be a contentious and exciting field in the next few years, and genetics should be able to offer some input.

A caveat

The spread of languages is a special case of cultural diffusion, or change. Unfortunately, the attempt to identify cultural change with the migration of people is now seen as old-fashioned in many archaeological circles. Instead, modern archaeologists stress indigenous reasons for the development of cultural attributes, or their borrowing from other cultures. The old school of diffusionism, which attempted to trace the expansion of particular cultures from a single place of origin, has fallen out of favour. However, the genetic results show that in some cases, this has clearly occurred. If genetic and cultural patterns overlap, as in the case of the eastern Dene-Caucasian languages, it is likely that there has been an ancient expansion of people carrying their culture with them. However, it is quite possible to expand a culture without a concomitant movement of people. This may have happened with the expansion of farming into north-western Europe.

As geneticists, we are limited by what we study. While we take history, archaeology and languages into account in our interpretations; our unique contribution is our ability to trace genealogy – actual biological relationships. Thus, we can find evidence to support human migration, as in the case of M17 and the steppe culture, as well as to refute it. Language is a good cultural attribute to study, since there are often written records. Even when there aren't, the relationships among languages can be examined systematically. Most cultural processes are not like this, making their interpretation more problematic.

One concept of race, popular until the mid-twentieth century, was that the different skin colours of people around the world reflected deep-seated biological differences. This was Carleton Coon's argument, and he used it (as well as skull shape and a few other characteristics) to divide humans into discrete population units. Earlier classifications had used cultural attributes as part of their racial definitions, as anthropologist Jonathon Marks has pointed out. Linnaeus, for example, had included 'obstinate, contented, free; paints himself with red lines' in his description of the American subspecies of *Homo sapiens*. Clearly, there was no biological basis to this – otherwise every Native American alive today would feel genetically compelled to paint his or her face. This archaic confounding of race and culture has had terrible consequences, most obviously during the heyday of eugenics. But, as we saw with the spread of languages, there are sometimes correlations between culture and genetics. Old-fashioned eugenicists may have taken this as evidence of a genetic *cause* for the cultural attribute, but in fact – as recent research shows – it is probably quite the opposite.

Sexual politics

The Karen people of northern Thailand and Burma are perhaps not as well known as their neighbours the Padaung, with their neck-extending rings of brass, but they are fascinating to ethnographers. This is because their social system runs counter to the pattern common in the vast majority of the world. Over 70 per cent of the world's societies practise something known as patrilocality. In this type of society, men control the wealth. Inheritance – and group membership – is passed through the male line. When two people marry, the wife goes to live with her husband and assumes a new identity in her husband's clan. The European custom of a wife changing her name to that of her husband traces its origin to this type of patrilocal behaviour.

One of the effects of patrilocal behaviour is that men tend to stay in one place while women are constantly immigrating into the family or clan. This may seem counter-intuitive – after all, don't men sow their wild oats more than women? – but it is the rule in most societies. The Karen, in contrast, do things differently. In Karen society, everything is turned upside-down. Women control the wealth, and group identity is passed through them to their daughters. In Karen marriages, men immigrate to the woman's village, taking over the care of her fields. Their society is what anthropologists term matrilocal, in that the women stay put and the men move. While the Karen may seem like ethnographic curiosities, in fact they have been instrumental in revealing the effect of culture on human genetic diversity. Like a bespoke experiment, they provide a social contrast to the prevailing pattern in human populations around the world.

We have used the Y-chromosome for most of our studies of human migration. This is because the Y shows greater differences in frequency between populations than most other genetic markers. As Dick Lewontin's analysis showed, most of the genetic diversity in the human species is found within populations, with a tiny fraction – 10 to 15 per cent – distinguishing between. For the Y, 30 to 40 per cent of the diversity is found between populations. Greater genetic contrast provides better resolution, which is why the Y is so good at tracing migrations.

When the Y was first studied as a marker of population affinity, one of the results that kept popping up again and again is that it connected people to a particular location. With a few DNA polymorphisms, it was possible to achieve incredible geographic resolution – there were even Y-chromosome polymorphisms that were limited to particular villages. If you imagine population genetics as a game of twenty questions, most genetic systems, including blood groups and mito-chondrial DNA, needed all twenty to identify even the coarsest pattern, such as which continent the individual came from. In contrast, the Y could typically identify subcontinental regions with a few questions. The observation, then, was that Y-chromosome lineages were geographically localized – they tended to define people as coming from a particular place. A fantastic tool for studying population movements, but the explanation for the pattern remained elusive.

In 1998 Mark Seielstad, then a graduate student working with Luca Cavalli-Sforza and Dick Lewontin, published a paper that proposed a solution to the Y mystery. Seielstad studied Y-chromosome markers in fourteen African populations, finding that the fraction of variation distinguishing between populations was much greater than that seen for other genetic markers. In a sample of European populations, the divergence between populations as a function of geographic distance increased at a much higher rate for the Y than for other genetic systems, such as mtDNA. Seielstad's interpretation of these two patterns was that women moved more than men, dispersing their mitochondrial lineages among neighbouring populations, producing a relatively homogeneous mtDNA distribution. The men, meanwhile, stayed at home – and their Y-chromosomes diverged independently in the different populations. The finding led Cavalli-Sforza to quip that Verdi was right when he wrote '*la donna e mobile*' (the woman moves).

Seielstad's publication created quite a stir, even attracting the attention of activists such as Gloria Steinem, who requested a copy. It seemed to undermine the ancient notion of peripatetic Lotharios wandering the globe, sowing their wild oats and dispersing their Y-chromosome lineages. What the activists failed to take into account, though, was that it actually reinforced the notion that women make a minor contribution to group identity. In a patrilocal society, it makes little difference who your mother was – it is your father who gives you your family or clan affiliation, and your inheritance. What Seielstad had found was that human culture has had a significant effect on the pattern of genetic variation in our species. Simple, local decisions about marriage and property, summed over hundreds of generations, had produced profound differences in the pattern of genetic variation on the male and female sides. Hindu castes show clear evidence of this pattern, with much greater Y-chromosome than mtDNA divergence between the castes, suggesting that women could move between castes while men were locked into theirs.

The real test of this theory, as Seielstad pointed out, was to examine the pattern of variation in matrilocal societies. The prediction is that these would show greater divergence for mtDNA, with the Y lineages tending to be homogenized among them. This was finally done in 2001, when Mark Stoneking and his colleagues published a study on the Karen, as well as a sample of patrilocal Thai tribes from the same area. They found Seielstad's predicted pattern of greater Y diversity in the Karen, providing strong evidence that patrilocality had produced the geographic clustering in Y-chromosome variation seen in most human societies.

While this helped to explain the localization of Y-chromosome lineages, it skirted round another odd observation. As we saw in Chapter 3, the coalescence time – the time elapsed since our common ancestors, Adam and Eve – is much more recent for the Y-chromosome than it is for mtDNA. Patrilocality can explain the high degree of Y divergence between populations, but the overall coalescence time should still be the same for Y and mtDNA. In effect, the Y pattern should be fragmented into many deeply divergent populations, all of which trace their ancestry to a single African man who lived around 150,000 years ago. Instead, we see many fairly divergent populations, all of which seem to coalesce to a common ancestor as soon as they are traced back to Africa: the data points to an African Adam who lived only a few thousand years before humans started to leave the continent. This result suggested another factor at work.

The rate of genetic drift – random changes in marker frequency due to small population size – depends on the actual size of the population, as we saw earlier. In large populations drift is negligible, while in small populations the effects of drift are significant. In the smallest populations, such as those first Beringeans who colonized the Americas, tiny population sizes can lead to a few lineages reaching frequencies of 100 per cent in a very short period of time. This is the explanation for why Native Americans are almost uniformly blood group O – types A and B were lost during their journey through the Siberian ice age.

This same pattern can be used to explain the recent dates for our

Y-chromosome ancestor. If there are fewer men than women in a population, then the rate at which Y-chromosome lineages are lost will be greater. But this can't be true, you might be saying – the birth ratio is 50 : 50. Surely there are the same number of men and women in every population? Surprisingly, while this is true in terms of numbers, it is not true for the number that pass on their genes by leaving offspring. In the genetic sense, those who don't reproduce don't count, and should be excluded from the equation. What we are interested in, then, is what is known as the *effective* population size – the number of breeding men and women. This is where we see the difference.

The likely explanation for why there is a greater rate of lineage loss for the Y-chromosome is that a few men tend to do most of the mating. Furthermore, their sons – who inherit their wealth and social standing – also tend to do most of the mating in the next generation. Carried through a few generations, this social quirk will produce exactly the sort of pattern we see for the Y-chromosome: a few lineages within populations, and different lineages in neighbouring populations. It will also produce a very recent coalescence time for the Y, since the lineages that would have allowed us to trace back to an Adam living 150,000 years ago were lost while our ancestors were still living in Africa. The definitive proof of this hypothesis will come only from careful studies of traditional societies, where the same social patterns have been practised for hundreds or thousands of years, but my prediction is that it will be confirmed by the data. As with the search for the language of Adam and Eve, the study of the effects of culture on human genetic variation promises to be one of the most exciting areas of enquiry in anthropology over the next few decades. Unfortunately, we may be racing against the clock, as we'll see in the next chapter.

Back to the sea

We've been through a tour of how culture, from the development of agriculture to local marriage patterns, has had an effect on human genetic diversity. We are now ready to re-evaluate the Hawaiians who were 'discovered' by Captain Cook in the late eighteenth century.

Where did they come from, and why had they conquered the Pacific in the last few thousand years?

The first question we can ask is whether there is a linguistic relationship among the Polynesian languages that suggests a source population. The answer is that there is. While Thor Heyerdahl favoured a South American origin for the Polynesians, their languages are more closely related to those spoken in south-east Asia. As early as the nineteenth century, scholars had linked the languages of Polynesia to those spoken in Taiwan (then Formosa) and Malaysia. Today, Taiwan is inhabited by Han-speaking Chinese, but prior to the seventeenth century it was home to aboriginal groups speaking completely different languages. All of these languages were united into one family, Malayo-Polynesian, which became known as Austronesian in the early twentieth century. So, there is clear linguistic data tracing from Hawaii back to Asia, rather than the Americas.

The overlap between the Austronesian languages and the spread of agriculture in east Asia is striking, and the theory which emerged for the peopling of Polynesia is that agriculturalists who had perfected the art of sailing simply hopped from island to island through south-east Asia, eventually heading into the open ocean. The 'Express Train' model, as it became known, predicted a close genetic link between aboriginal Taiwanese and the Polynesians. MtDNA seemed to support this model, although its resolution – as we have seen elsewhere – is often limited. Recent results from the Y-chromosome, though, have suggested that the theory needs to be modified.

The pattern seen for the island south-east Asians is that, while agriculturalists of (ultimately) Chinese origin did have a significant impact on the gene pool, there are a substantial number of indigenous lineages (particularly M130) found throughout Indonesia and Melanesia. These are also present at high frequency in the Polynesians. What this suggests is that after agriculture was introduced to island south-east Asia, it went through a maturation phase as it was adapted to local crops that were better suited to the environment there. Instead of flying past on their express train, the agriculturalists dawdled and dabbled, gradually adapting their culture to its new home. Archaeologist Peter Bellwood has pointed out that the crop yield of Chinese rice strains drops significantly if they are planted near the equator, since they need the variation in day length found only outside the tropics in order to mature. These sorts of pressures would have encouraged agriculture to change as it passed through south-east Asia, in some cases replacing millet and rice with other crops. The Polynesian taro root, ubiquitous throughout the Pacific and used to make Hawaiian *poi*, reflects this change. The genes also show evidence of a sojourn in south-east Asia before heading out to sea.

The answer to our question of timing, then, can be found in the maturation phase of agriculture. It was only after a fully mature tropical variant of agriculture had taken root that the proto-Polynesians were able to set sail for undiscovered lands. They took with them their crops, confident in their ability to survive wherever they came ashore. Hunter-gatherers would never have been able to make this leap into the unknown ocean – repeatedly – because they had no idea what lay beyond the horizon. The Polynesians, though, as inheritors of a well-adapted agricultural tradition, were in control of their own destinies. They may have been encouraged to set sail by an expanding population at home (another consequence of agriculture), but their unique solution was only possible because they had the *choice* of sailing into the unknown. And it was the pursuit of an ever-increasing spectrum of choices that would produce the final Big Bang of human evolutionary history.

Figure 9 Genealogical tree showing the relationship among the Y-chromosome markers discussed in the text. All trace their descent from M168, who lived in Africa.

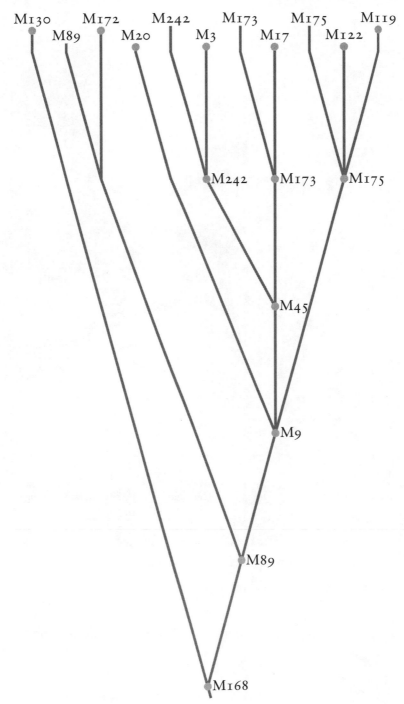

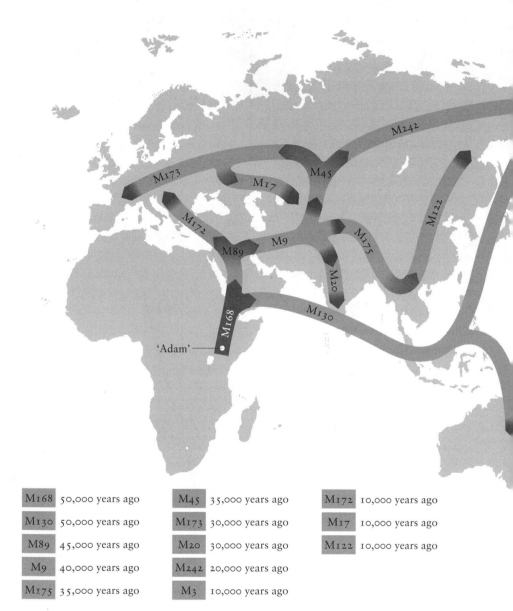

M168	50,000 years ago	M45	35,000 years ago	M172	10,000 years ago
M130	50,000 years ago	M173	30,000 years ago	M17	10,000 years ago
M89	45,000 years ago	M20	30,000 years ago	M122	10,000 years ago
M9	40,000 years ago	M242	20,000 years ago		
M175	35,000 years ago	M3	10,000 years ago		

Figure 10 The spread of Y-chromosome lineages around the world.

9
The Final Big Bang

If you know your history, then you know where you're coming from.
Bob Marley, 'Buffalo Soldier'

A couple of years ago, I was asked to perform a genetic analysis as part of a television programme. The goal was to show, using genetic data, that all humans trace back to a recent African ancestor. Initially I was hesitant, since it would involve revealing personal genetic results in front of a television camera for the whole world to see. But after being reassured by the producers and the people who donated samples, I went ahead with the analysis. Four men living in London volunteered to have their Y-chromosomes tested, and I analysed the markers we have encountered in this book – M168, M130 and so on. When the work was complete, the data showed the expected pattern for three of the four men. The man with Irish/Scottish ancestry had a Y-chromosome defined by M173, the highest frequency Y lineage in north-western Europe. The Japanese man had M122, in common with around 20 per cent of his fellow countrymen. The Pakistani had the M89 lineage, found throughout the Middle East and central Asia. The final man, though, had a surprising pattern. An Afro-Caribbean, he was hoping for a genetic link to the Zulus of southern Africa, with whom he felt a strong cultural bond. The DNA revealed a more complicated story.

This man turned out to have an M173 Y-chromosome, the canonical European lineage. M173 has never been identified in hundreds of samples from indigenous sub-Saharan Africans, so the obvious question was how did he come to have such an anomalous result? The other, non-Y markers we tested revealed him to be otherwise genetically African – including the presence of a marker I had first identified in a Zulu man in the mid-1990s. Clearly, the Y was telling a different story – one that helps to illustrate the main theme of this chapter.

The reason our Afro-Caribbean man had a European Y-chromosome was that, at some point in the past, one of his male ancestors must have had a European father. Given his family history, it is likely that this occurred when his family lived in the Caribbean, during the era of slavery. Clearly, knowing the history of recent migration was critical to interpreting this result. Once the circumstances were recognized, it became very easy to reconcile the Y data with the rest of the genetic story, giving us a glimpse into his complicated family tree.

Was this a unique case? Absolutely not. Among African–Americans, as many as 30 per cent of Y-chromosome lineages appear to be European in origin. The slave-trading era has left a distinct pattern in the DNA of people of African descent living outside Africa. They are not unique in having a mixed ancestry, though. In the past 500 years – encompassing the European Age of Exploration and the Industrial Revolution – humans have become far more mobile than ever before. Today, the descendants of those first modern humans to wander into Eurasia zigzag around the planet at a pace that would leave our Upper Palaeolithic ancestors breathless. The final Big Bang in human evolutionary history – which could be called the Mobility Revolution – has given rise to the era of globalization. While the cultural and economic consequences of living in a 'global village' are debated by businesspeople and policymakers, and the ecological fallout is seen in the accelerating loss of biodiversity, the genetic effects of the latest Bang are perhaps less clear.

A linguistic thread

Much of my work as a geneticist has focused on deciphering the relationships among the people living in central Asia. The former Soviet states of Uzbekistan, Kazakstan, Kyrgyzstan and their neighbours were locked away from most Western scientists during the Soviet era, and when they opened up in the early 1990s I jumped at the chance to go there. Sampling of the world's genetic diversity had, until that time, concentrated mostly on Europe, east Asia (particularly China and Japan), South Africa and North America. Central Asia was pretty much unknown – a 'black box' in the genetic patterns of the world.

I first visited in the summer of 1996, and since then my work has taken me there several times. I have driven there from London in a Land Rover, flown in on shaky Soviet-era planes and walked across remote borders carrying bags of genetic sampling equipment. One of my most memorable trips, though, was when I visited Tajikistan in August 2000. Working with local scientists and physicians, our aim was to take blood samples from several of the ethnic groups living in the mountainous regions of the country. One of these was the Yagnob.

The Yagnob are a direct link back to the days of the Silk Road. Their language, Yagnobi, is a direct descendant of Sogdian, which was once the lingua franca of the Silk Road – in much the same way that English is the language of commerce today. In the mid-first millennium AD, Sogdian was spoken in trading centres across central Asia, from Persia to China. After the Muslim conquests of the seventh to ninth centuries its use declined, and by the twentieth century all of the dialects were extinct – except one. The Yagnob people, living in a few isolated villages in the remote Zerafshan Valley of northern Tajikistan, still speak this ancient language, a 1,500-year-old linguistic artefact. Our plan was to visit them and explain our project, hoping that they would want to participate in the research project to trace their history using the signals in their DNA.

The trek up to the Yagnob villages from the Tajik capital, Dushanbe, involved crossing a pass that had only recently been retaken by government forces in Tajikistan's long and bloody civil war. After passing through several checkpoints guarded by Kalashnikov-toting soldiers and descending to the series of parallel valleys on the other side, we found a dirt road leading eastward, alongside the Zerafshan River. Several hours later, after pushing our old Soviet van through the rough patches, we reached a small *kishlak*, or village. Expectantly we hopped out and asked to speak to the local 'head man'. We explained the project, sipping tea while the old man pondered what we had said. Eventually, he told us that we had made the trip in vain.

The Yagnob had lived here for generations, he explained – perhaps even since the days of the Silk Road. But in the 1960s, drought had led the Soviets to resettle them in villages in the lowlands. Also, in the late 1980s there had been an earthquake, and many of those who remained had moved to Dushanbe. Now it was actually very difficult to find Yagnob living in their ancient land. You might find cab drivers or cleaners in the capital who came from this region, but – except for a remote village several days' hike into the mountains – the Yagnob had largely decamped from their ancient homeland. Disappointed, we thanked him and left. After another couple of days' searching we managed to discover one Yagnob village, and the locals were very happy to help us with our work, but in the end we collected more samples from this ancient population by trolling around the capital. Our effort to find an isolated remnant of the Silk Road had nearly failed.

What the old man in Tajikistan had explained to us actually happens every day, all over the world. The Yagnob are not an unusual case – in fact, quite the opposite. It is a fact of modern life that villages are gobbled up by growing cities, their residents thrown into a mix of languages and ethnicities that becomes ever more complex as the city expands. And while some societies tolerate diversity, many view it as an impediment to unity. It is often shunned by governments keen on cultivating cultural harmony – especially in newly created states striving for a sense of identity. To understand why, we need to take a closer look at the model for statehood developed in nineteenth-century Europe.

Withering tongues

When you visit France today, it is difficult not to be impressed by the people's love for their language. The Académie Française, official guardian of the national language, monitors spoken and written French like a hawk, helping to whip it into shape in the face of foreign influences. Yet just 150 years ago – roughly six generations – fewer than half the people living in France actually spoke French. Most spoke their local dialects and languages. In Italy around the same time, less than 10 per cent of the population was estimated to have spoken Italian. Austrian chancellor Clemens von Metternich quipped at the time that Italy was less a nation than a 'geographical expression' – clearly true if language was a factor.

Nineteenth-century Europe was a swirl of new ideas and movements. Romanticism, realism, industrialization, colonial expansion – all would make significant contributions to the development of our 'modern' worldview. One of the most important manifestations of the new thinking was the rise of nationalism, which was to create the modern political map of Europe – and have far-reaching effects on the rest of the world.

Before the nineteenth century, Europe was divided up into separate fiefdoms, kingdoms and duchies. Life was much more 'local' than it is today. People's allegiance was to provincial rulers, and their lives revolved around regional events. This was reflected in their marriage patterns, which tended to be highly localized. The distance between spouses' birthplaces was only a few kilometres throughout most of European history, leading to high levels of consanguinity, or intra-family marriages. These regional characteristics extended to language as well. For instance, while modern France has one official language, religiously guarded by the Académie, in the late eighteenth century there were many provincial languages tracing their existence back hundreds or thousands of years. Basque, Breton, Occitan, Corsican, Alsatian – all were distinct linguistic entities. Breton, for instance, is a Celtic language more closely related to Welsh and Gaelic than to French, despite the fact that it was the language of Brittany on the north coast of France. The speakers of these regional languages saw themselves as possessing unique identities – ethnicities, if you will – which were to be subsumed in the process of creating the French state.

As nationalism took hold in Europe, language was used by the newly unified states to create a sense of national identity. Governments sought cultural unity by favouring one language over the others. From the eighteenth century onward English was the primary literary and governmental language in the United Kingdom, but many people living in the UK spoke languages only distantly related to English. The effect was an expansion in the number of native English speakers at the expense of the Celtic languages. The Celtic Manx language (known locally as Ghailckagh), native to the Isle of Man, had 12,000 native speakers in 1874, but only 4,000 at the turn of the twentieth century. The last first-language Manx speaker died in 1974, and today it is kept alive only as a kind of living fossil by a few hundred aficionados.

During the nineteenth century compulsory schooling in the national language, as well as national military service, helped to spread the chosen tongue, and within a few generations the process was nearly complete. Nationhood had been transmogrified into monolingualism. One of the best examples of the entanglement of language and nationhood comes from Germany. The Grimm brothers, Jacob and Wilhelm, are famous for compiling the fairytales that most European children hear during childhood. It is perhaps not as widely known that Jacob was also an accomplished linguist, who defined the rule for the sound changes that occurred during the evolution of the Germanic languages – for instance, when a b in an ancestral Indo-European word became a p in German, and so on. The Grimms' work was done, at least in part, in order to derive a sense of unity for the German-speaking peoples. In the case of the linguistic studies, it was an effort to define and codify the unity and history of the Germanic languages, as part of the creation of a national language standard. The fairytales, on the other hand, were an effort to record the folk culture of the Germans, in order to preserve and mould their national identity. Germany was in the process of becoming 'German', and the Grimms were among the intellectual architects of the new nation.

The identification of history with language was developed during this period of European nationalism, but it was simply a formal statement that languages tend to define cultures, and cultures are intimately tied to their languages. The reason for this is the length of time it takes to 'create' a language – around 500–1,000 years to develop something that is distinct from its sibling tongues. The Romance languages, for instance, have been diverging from each other for around 1,500 years, tracing their origins back to the days when Latin was the language of the Roman Empire. Today French, Spanish, Italian, Romanian, Catalan and Romansch (spoken in the Swiss canton of Graubunden) are all related through a common ancestry to the language of the Romans.

Other languages, such as Basque, have been distinct from their surrounding languages for much longer. But in each case, a language represents the end result of many years of cultural isolation.

When languages are lost, then, we lose a snapshot of one part of our history. If the Basque language went extinct, we would lose the only remaining link back to the pre-Indo-European languages of Europe. If the estimated 2,000 Yagnobi speakers in Tajikistan are completely integrated into the Tajik-speaking majority, and their children stop learning Yagnobi, then we will have lost this living connection to the time of the Silk Road. In every case of language death, we lose a part of our cultural history. Particularly when the language in question has not been studied and recorded – which is the case for most of the world's languages – we have lost an irretrievable snapshot of our past.

Today, the fifteen most common languages, in terms of number of speakers, are spoken by half the world's population. Some of these languages (English, Spanish and Arabic among them) were spread through colonialism. Others have increased in number through population growth, spurred on by agriculture – Chinese and Hindi being the two best examples. Even in these cases, though, the creation of a national language has contributed to their success. What is clear is that a few languages are becoming much more widespread. The top 100 languages are spoken by 90 per cent of the people in the world – despite the fact that linguists recognize over 6,000 distinct tongues. Clearly, most are spoken by only a few people.

The future of most of these languages is uncertain at best. Through the same processes that have reduced the number of Yagnobi and Manx speakers, most languages are headed for extinction. Most of these doomed languages are spoken by small populations that have been absorbed into or dispersed by larger groups. The Yaghan language – spoken by Darwin's Fuegians whom we heard about in Chapter 1 – is probably already extinct, a victim of European colonialism. Linguists Daniel Nettles and Suzanne Romaine estimate that over half of the world's languages could be extinct by the end of this century – a rate that equates to the loss of one language every two weeks. There were estimated to have been 15,000 languages spoken around the world in 1500, so we have already lost more than half of the linguistic diversity that once existed.

But, you may be thinking, the focus of this book has been on what our genome tells us about our history. Why, then, should we care about the rise of nationalism and the loss of languages? Because, as we saw in the previous chapter, languages are often correlated with genetic patterns. In that case, what does the loss of linguistic diversity reveal about the current state of our genomes – and their future?

The global melting pot

As we have seen, only a tiny fraction of the genetic diversity in the human species distinguishes populations from each other – the vast majority of variation is found within a single population. There are two reasons for this. The first is that we are a relatively young species. Around 50,000 years ago – only 2,000 generations – our ancestors all lived in Africa. Given that mutations happen only infrequently, and that it takes a while for them to increase in frequency to a point where they can be sampled in a population, it is likely that most of the diversity we see now already existed in this ancestral African population. This is particularly true for polymorphisms other than those on the Y-chromosome. Most of these other polymorphisms appear to be quite old, consistent with the fact that they were present in the ancestral population before our journey out of Africa.

Furthermore, human 'races' seem to have very recent origins. For the most part, physical traits that distinguish modern human geographic groups only appear in the fossil record within the past 30,000 years. Most older fossil Africans, Asians and Europeans are very similar to each other. While we know nothing about our ancestors' skin colour, hair type or other surface features, the evidence from bones suggests that our concept of race is actually a very recent phenomenon. It was probably the fragmentation of human groups as a result of the last ice age that produced the distinct 'racial' morphologies we see in modern humans – not hundreds of thousands of years of separate evolution, as Carleton Coon and others had argued. For example, sinodonty – the distinctive tooth pattern common to north-east Asia and the Americas – first appeared in the fossil record less than 30,000 years ago. Before then Asian teeth were very similar to those seen elsewhere in the world.

The other reason for genetic uniformity among human populations is that humans are mobile, and groups have intermixed throughout their history. When this happens, their patterns of genetic variation become dispersed throughout the mixed population. So, even in cases where the genetic markers arose after modern humans migrated out of Africa – like most of the markers we have followed on the Y-chromosome – they will still be widely distributed as a result of subsequent mixing.

The dynamics of language extinction indicate that human mixing is now accelerating. Languages seem to die out primarily through the incorporation of small, previously isolated populations into a larger, dominant population – in the same way that Manx was subsumed into English-speaking Britain. It is rarely the case that a minority population actually dies – rather, it is simply incorporated into the majority. But is there any real data on the rate at which this is happening?

The answer is yes. Most developed countries have a national census, where the people living in the country are counted and subdivided into demographic units. The reasons for this may be pragmatic – apportioning political representation or government funds, for instance – but the data also reveals deeper truths about society. Perhaps the best-known census is that held every ten years in the United States, the most recent of which was in 2000. Aside from showing that the population of the USA was 281.4 million, an increase of 13 per cent over that in 1990, it also detailed a changing ethnic landscape. For the first time in 2000, people were able to subdivide their ethnicity accurately. The number of racial categories expanded from five to sixty-three, and for the first time combinations of minority groups could be reported.

In total, 6.8 million people described themselves as a mixture of 'white' and a 'minority' group. Of course, this ignores the mixing that has created the category of white, which could mean anything from Irish to Lebanese to Moroccan. As we have seen in previous chapters, this mix alone encompasses a wide range of populations and markers. In terms of the official 'mixed' category, many people who have mixed ancestry actually consider themselves to be one race to the exclusion of the other, suggesting that the true number of mixed Americans is actually much higher than what was actually reported. For instance, surveys carried out by the US Census Bureau show that, while only 25 per cent of black/white mixed respondents considered themselves white, around half of white/Asian and white/Hispanic respondents, and 81 per cent of those with mixed white/Native American ancestry thought of themselves as white. One of the results of the 2000 census is that it became clear that America is far more of a melting pot than it may have imagined.

The golfer Tiger Woods may be more indicative of the face of today's United States than many realize. Woods, who claims African–American, European and south-east Asian ancestry, falls into that ever-increasing group of people who would find it difficult to describe their ethnicity in simple terms. Even those who give themselves a single classification, such as African–American, often have substantial admixture from other groups. This was actually one of the criticisms levelled at the first scientific publication on mitochondrial Eve in 1987. Because Cann, Stoneking and Wilson had sampled African–Americans living in the San Francisco Bay area as their representative 'African' population, critics noted that it was possible that the deepest lineages in their analysis – those indicating an African origin – may actually have been non-African. It was only in their second paper, in 1991, that Africans were included, indicating that the conclusions of the original publication were valid.

Tiger Woods is in many ways a person who could only have been born in the twentieth century. His complex web of ancestors, originating on opposite sides of the world, could have encountered each other in the United States only within the past 100 years. But Mr Woods is merely an obvious example of a phenomenon that has been ongoing for the past few centuries, resulting in the jostling together of people who, historically speaking, would never have met. Coupled with changing social attitudes towards race, people today are far more likely than their ancestors to have children of mixed ethnic backgrounds. While this is certainly a good thing socially, leading to the breakdown of racial stereotypes, it does mean that our genetic identities are becoming ever more closely entwined. Admixture is destroying the old, regional patterns of genetic diversity, replacing them with cosmopolitan melting pots of markers. It is likely that sampling 100 people in a nightclub in New York's East Village would reveal every single one of the markers we have discussed in this book, all present in one small, potentially interbreeding population. The implications of this sort of mixing for our studies of genetic history will be the final stop on our journey.

A closing window

The third Big Bang of human history has led us into a new genetic landscape. The patchwork quilt of diversity that has distinguished us since human populations started to diverge around 50,000 years ago is now re-assorting itself, blending together in combinations that would never have been possible before. While the genetic markers themselves will not be lost, the context in which they arose may soon be gone. And although we can trace the genetic relationship among lineages as easily in a sample from our New York nightclub as from isolated groups all over the world, the result will have little meaning. This is because we cannot place the genetic analysis in a geographic location. Our coastal voyage to Australia, for instance, relies on the distribution of the oldest M130 chromosomes being limited to the southern part of Eurasia, and their absence from the Middle East. It is only by sampling indigenous people who have lived in these places for a long period of time – in this case, ideally, 50,000 years – that we can hope to infer the genetic makeup of their ancestors. Ancient, local populations are key, and the less admixture they have had the better. These are exactly the groups that are now being lost. Taking languages as a litmus test, isolated communities are being engulfed at an increasing rate. Moreover, because of the nature of modern industrial life, members of these communities are increasingly moving to cities, where their markers will enter the vast, swirling melting pot of cosmopolitan diversity. Unfortunately, when this happens the unique story they have to tell will be lost.

Some minority populations are rediscovering a sense of identity, fighting against the advancing wave of global culture. European activists such as the Basque ETA, French farmers who bomb McDonald's restaurants and the ranks of anti-globalization protestors at economic summits – all are a sign of the growing realization that cultural identity is being lost on a massive scale. Ultimately, though, their methods are too extreme to achieve widespread support. And for most indigenous people, the rewards of becoming part of the global village are simply too enticing to be ignored. Decisions to leave ancient villages usually come down to personal choices – a perception that opportunity is better elsewhere, or that it has disappeared at home. In the end, because they cannot limit personal choices, it is a battle the activists are doomed to lose.

The story in this book could only have been told now. However, it is merely the outline of a much more detailed narrative, the whole of which will take many more years of research to decipher. We may have a view of the forest, but we still know very little about the trees. With the realization that their cultural identity is being eroded, though, many indigenous populations are now refusing to participate in scientific studies. A history of colonial exploitation, with incidents such as the horrendous medical experiments inflicted on the Australian Aborigines in the mid-twentieth century, has understandably led many indigenous people to be wary of scientists. Activists are also reasserting ancient taboos on ancestor disinterment, asking for archaeological material to be returned for proper burial. These cultural taboos can, and do, extend to giving samples for genetic studies. In a way, we are trying to excavate the past from the blood of people living in the present – an activity that can be interpreted as voyeuristic (or worse). A desire for cultural privacy, perhaps combined with the suspicion that the scientific results may not agree with their own beliefs, is leading more and more indigenous groups to choose not to participate. Scientists have a responsibility to explain the relevance of their work to the people they hope to study, in order for their participation to become what it really is – a collaborative research effort. Only then we can regain some of the trust we have lost.

Today we are in many ways the same Palaeolithic species that left Africa only 2,000 generations ago, with the same drives and foibles. It is ironic that the final Big Bang of human history, which has given us the tools to 'read' the greatest history book ever written – the one hidden in our DNA – has also created a cultural context where it is becoming increasingly difficult to carry out this work. The genetic data we have glimpsed shows unequivocally that our species has a single, shared history. Each of us is carrying a unique chapter, locked away inside our genome, and we owe it to ourselves and to our descendants to discover what it is. Since our ancestors came down from the trees, we have used our intellect to explore outward and extrapolate into the future. Over the past few thousand years we have changed our world – and our place in it – for ever. With the development of agriculture, and the cultural chain reaction it ignited, we gained the power to choose our own evolutionary trajectory. With this power, though, came increased responsibility. One responsibility that we neglect at our peril is that of self-discovery. Once the document of our journey has been lost it will, like the footprints of our ancestors as they left Africa to colonize the globe, be gone for ever.

Acknowledgements

I have benefited enormously from the help and insight of many colleagues, who have provided data, interpretations and counter-arguments for the many theses I pursue in the book. Foremost among them is Peter Underhill, whose careful work on the population genetics of the Y-chromosome has allowed me to tell this story. It was Peter and his colleagues at Stanford who discovered most of the markers discussed in this book, and the field owes him a debt of gratitude. I have also learned a great deal from my work with Li Jin, a fountain of knowledge on the population history of east Asia, and from interactions with my Oxford colleagues Walter Bodmer, Tatiana Zerjal, and Chris Tyler-Smith, who have challenged me on various genetic details and always make for very stimulating company. Nadira Yulda-sheva and Ruslan Ruzibakiev have been friends and co-workers during countless months of sample collecting in remote parts of Asia, and throughout the years of labwork that followed – *bolshoi spasibe*. Merritt Ruhlen and Richard Klein were happy to discuss their work on linguistics and palaeoanthropology, respectively, which was invaluable. Thanks also to Lluis Quintana-Murci, Matthias Krings and Mark Seielstad for in-depth explanations of their work over long, boozy meals in Paris, London and Boston – the hangovers were worth it. My colleagues at Tigress Productions in London, who believed in this project during the long television commissioning process, have created a wonderful film – thanks to Jeremy, Justine, Clive, David, Ceri, Jackie, Aidan, and Martin. We were lucky to have a great producer, Jennifer Beamish, whose sharp mind acted as a perfect sounding-board for many of the ideas in this book. A special thanks to my editor at Penguin, Stefan McGrath, whose enthusiasm for this project has never waned, and who was able to make deft use of both carrot and stick in order to get me to finish the book on time during our long filming schedule – I owe you a few beers. And finally, apologies to my wife, Trendell, and to my daughters, Margot and Sasha, for my long absences during this project. Even when I was home, I was often preoccupied – thanks for bearing with me.

The best overall summary of human genetic patterns, and their relationship to prehistory, is *The History and Geography of Human Genes* by Luca Cavalli-Sforza, Paolo Menozzi and Alberto Piazza (Princeton University Press, 1994). This extraordinary volume summarizes over thirty years of work on classical genetic polymorphisms in human populations, and is the best single-volume reference available on the more technical aspects of much of the material in this book. Cavalli-Sforza's more approachable *Genes, Peoples and Languages* (Penguin, London, 2000) presents some of his ground-breaking work for a general audience.

Three other books stand out as indispensable introductions to human prehistory, more from the perspective of stones and bones than DNA: Richard Klein's *The Human Career* (2nd edition, University of Chicago Press, 1999), Brian Fagan's *People of the Earth* (8th edition, HarperCollins, New York, 1995) and Chris Stringer and Robin McKie's *African Exodus* (Pimlico, London, 1996). All three take a very broad view of human prehistory, and Klein's book in particular is tied together with a persuasive argument (which I also pursue in the present volume) that the intellectual leap that took place in Africa around 50,000 years ago allowed our species to colonize the rest of the earth.

1 The Diverse Ape

The best English translation of Herodotus' *History* I have found is that by David Grene (University of Chicago Press, 1987). It is written in a vernacular style that manages to communicate the excitement of the Greek historian's world in a fresh way – nearly 2,500 years after it was written.

Darwin's *Beagle* journal has been published and reprinted many times – the version I have used is *The Voyage of the Beagle* (Modern Library, New York, 2001), with an interesting introduction by Steve Jones. Many of the biographical details of Darwin's life came from Janet Browne's wonderfully readable *Charles Darwin: Voyaging* (Alfred A. Knopf, New York, 1995) – the first in a planned two-volume definitive biography. Darwin's *On the Origin of Species* (Harvard University Press, Cambridge, MA, 1964) and *The Descent of Man* (Princeton University Press, 1981) are so well known that they need no introduction.

Carelton Coon's work was summarized in his two influential books *The Origin of Races* (Alfred A. Knopf, New York, 1962) and *The Living Races of Man* (Alfred A. Knopf, New York, 1965). Daniel Kevles's excellent summary of the perversion of a naïve ideal can be found in his *In the Name of Eugenics* (Alfred A. Knopf, New York, 1985), and additional material is covered in Stephen Jay Gould's *The Mismeasure of Man* (W. W. Norton, New York, 1981) and Jonathan Marks's *Human Biodiversity* (Aldine de Gruyter, New York, 1995).

2 *E pluribus unum*

The title of this chapter – Latin for 'out of many, one' – is the motto on the Great Seal of the United States of America, found on all US coins.

The history of blood group studies and their application to human population genetics has been summarized in Arthur Mourant's seminal book *The Distribution of the Human Blood Groups* (Blackwell, Oxford, 1954). Much of my examination of Lewontin's work comes from many hours spent discussing genetics and human diversity with him, but many of his ideas are explained in *The Genetic Basis of Evolutionary Change* (Columbia University Press, New York, 1974) and *Human Diversity* (Scientific American Press, New York, 1982). His original paper analysing human genetic variation was published in the *Journal of Evolutionary Biology* (6: 381–98, 1972) – one of the most important twentieth-century publications in the field of human genetics.

Theodosius Dobzhansky's *Genetics and the Origin of Species* (Columbia University Press, New York, 1982) and Motoo Kimura's *The Neutral Theory of Molecular Evolution* (Cambridge University Press, 1983) are good summaries of these scientists' contributions to population genetics.

Cavalli-Sforza's work is summarized in *The History and Geography of Human Genes* and *Genes, Peoples and Languages* (see above). The original papers describing human population trees were published by Edwards and Cavalli-Sforza in V. E. Heywood and J. McNeill (eds.), *Phenetic and Phylogenetic Classification* (The Systematics Association, London, 1964, pp. 67–76), Cavalli-Sforza and Edwards in *Proceedings of the 11th International Congress of Genetics* (2: 923–33, 1964), and Cavalli-Sforza, Barrai and Edwards in *Cold Spring Harbor Symposium on Quantitative Biology* (29: 9–20, 1964). Cavalli-Sforza and Bodmer's *The Genetics of Human Populations* (W. H. Freeman, San Francisco, 1971) is a classic textbook – luckily, it has recently been reprinted by Dover after being unavailable for many years.

Parsimony is discussed in much greater detail in Elliot Sober (ed.), *Conceptual Issues in Evolutionary Biology* (MIT Press, Cambridge, MA, 1984), and in Arnold Kluge's contribution to T. Duncan and T. F. Stuessy (eds.), *Cladistics: Perspectives on the Reconstruction of Evolutionary History* (Columbia University Press, New York, 1984, pp. 24–38).

Zuckerkandl and Pauling's work on the use of molecules to infer evolutionary history was published in several journal articles during the early 1960s; perhaps the best summaries are in M. Kasha and B. Pullman (eds.), *Horizons in Biochemistry* (Academic Press, New York, 1962, pp. 189–225) and *Journal of Theoretical Biology* (8: 357–66, 1965). Rebecca Cann, Mark Stoneking and Allan Wilson's work on mitochondrial Eve was published in *Nature* (325: 31–6, 1987), and followed up by Vigilant et al. in *Science* (253: 1503–7, 1991). The analysis of complete mtDNA sequences (showing an unequivocally African origin for world mtDNA lineages) was published in *Nature* by Ingman et al. (408: 708–13, 2000).

An excellent historical summary of early palaeoanthropological work is Eric Trinkaus and Pat Shipman's *The Neanderthals* (Vintage, New York, 1992). Additional material can be found in the books by Brian Fagan and Richard Klein cited above, as well as in *Java Man* (Little, Brown, London, 2000) by Garniss Curtis, Carl Swisher and Roger Lewin, and in Robin McKie's *Ape Man* (BBC, London, 2000).

3 Eve's Mate

Other DNA studies supporting an African origin for modern humans were published by Wainscoat et al. (*Nature* 319: 491–3, 1986), Tishkoff et al. (*Science* 271: 1380–7, 1996) and Jin et al. (*Proceedings of the National Academy of Sciences USA* 96: 3796–800, 1999). There are others, analysing different regions of the genome, but all show essentially the same pattern – greater genetic diversity within Africa.

Two good technical reviews on the structure and evolution of the Y-chromosome are Jobling and Tyler-Smith's in *Trends in Genetics* (11: 449–56, 1995) and Lahn et al.'s in *Nature Reviews Genetics* (2: 207–16, 2001). Early papers on Y-chromosome variation were those by Casanova et al. (*Science* 230: 1403–6, 1985), Lucotte and Ngo (*Nucleic Acids Research* 13: 82–5, 1985), Dorit et al. (*Science* 268: 1183–5, 1995) and Hammer (*Science* 378: 376–8, 1995). DHPLC and its application to Y-chromosome population genetics is discussed in Underhill et al. (*Genome Research* 7: 996–1005, 1997). The paper by Underhill et al. dating Adam to 59,000 years ago was published in *Nature Genetics* (26: 358–61, 2000).

4 Coasting Away

Bruce Chatwin's *The Songlines* (Vintage, London, 1987) gives a general introduction to aboriginal culture. Other good sources on Australian prehistory are A. W. Reed's *Aboriginal Myths, Legends & Fables* (Reed New Holland, Sydney, 1993), Kleinert and Neale's *Oxford Companion to Aboriginal Art and Culture* (Oxford University Press, 2000) and Tim Flannery's *The Future Eaters* (Reed New Holland, Sydney, 1994). The archaeology and geology of Lake Mungo is described in Allan Fox's *Mungo National Park* (Beaten Track Press, Yarralumla, 1997). The dates for the Lake Mungo human remains are currently being revised, and I benefited greatly from my discussions with archaeologist Doug Williams, Executive Officer of the Willandra Lakes World Heritage Area, based in Buronga, New South Wales.

A good introduction to African geography and climate is Lewis and Berry's *African Environments and Resources* (Unwin Hyman, Boston, 1988). Robert Walter and colleagues' research on African coastal dwellers was reported in *Nature* (405: 65–9, 2000). The mtDNA evidence for a coastal exodus from Africa was published by Lluis Quintana-Murci in *Nature Genetics* (23: 437–41, 1999). The Y-chromosome data on the distribution of M130 (also known as RPS4YT) is taken from three publications: Kayser et al. (*American Journal of Human Genetics* 68: 173–90, 2001), Underhill et al. (*Annals of Human Genetics* 65: 43–62, 2001) and Wells et al. (*Proceedings of the National Academy of Sciences USA* 98: 1044–9, 2001). The archaeological evidence (or lack thereof) for an Upper Palaeolithic coastal migration is presented in Peter Bellwood's *Prehistory of the Indo-Malaysian Archipelago* (University of Hawaii Press, Honolulu, 1997) and Gregory Possehl's and Charles Higham's articles on south and south-east Asian prehistory, respectively, in *The Oxford Companion to Archaeology* (Oxford University Press, 1996, pp. 52–7).

The scenario proposed in this chapter, of populations beachcombing their way to Australia, is similar to one advanced by Jonathan Kingdon in *Self-made Man and His Undoing* (Simon & Schuster, New York, 1993).

5 Leaps and Bounds

The term Great Leap Forward was first applied to the study of human prehistory by Jared Diamond in *The Rise and Fall of the Third Chimpanzee* (Vintage, London, 1991) – a fascinating summary of human prehistory. Good sources on the origin of language include Steven Mithen's *The Prehistory of the Mind* (Phoenix, London, 1996), Steven Pinker's *The Language Instinct* (William Morrow, New York, 1994) and Parker and McKinney's *Origins of Intelligence* (Johns Hopkins University Press, Baltimore, 1999). William Calvin's *A Brain For All Seasons* (University of Chicago Press, 2002) discusses the impact of climate change on human brain evolution. Thomas Keenan's *An Introduction to Child Development* (Sage, London, 2002) is a good general overview of this very complicated subject.

Henry Harpending and colleagues' work on the inference of human population expansions from mitochondrial DNA data is presented in a paper in *Human Biology* (66: 761–75, 1994). Much of the information on climate change in Africa and the fossil record in the Middle East was taken from Richard Klein and Stringer and McKie's books (see above), as well as that of John Gowlett (*Ascent to Civilization*, Alfred A. Knopf, New York, 1984).

6 The Main Line

The ordering of the Y-chromosome markers discussed in this chapter, and their implications for human migration, appear in Underhill et al.'s 2000 *Nature Genetics* and *Annals of Human Genetics* papers (see above). The spread of Y-chromosome lineages along the Eurasian steppe belt is discussed in the Wells et al. *Proceedings of the National Academy of Sciences* paper. A good overview of the central Asian fossil record is Dani and Masson's *History of the Civilizations of Central Asia*, Volume 1 (UNESCO, Paris, 1992). Lewis Binford's work on the importance of scavenging in the early human diet has been presented in many publications, one good example being in *Journal of Anthropological Archaeology* (4: 292–327, 1985). Cavalli-

Sforza's work on Chinese populations is discussed in *The History and Geography of Human Genes* (see above).

7 Blood from a Stone

James Riordan's *The Sun Maiden and the Crescent Moon: Siberian Folk Tales* (Interlink Books, New York, 1989) is a great introduction to the stories of Siberia's native peoples. A good overview of Upper Palaeolithic cave art is Paul Bahn's *Journey Through the Ice Age* (Seven Dials, London, 1997) – beautifully illustrated with Jean Vertut's photography.

The first Neanderthal sequence was published by Matthias Krings and his colleagues in *Cell* (90: 19–30, 1997) – truly a landmark paper in the study of human origins. The dating of M173, the major western-European Y-chromosome lineage, is given by Semino et al. in *Science* (290: 1155–9, 2000). Ezra Zubrow's modelling of Neanderthal demographic patterns is in Stringer and Mellars (eds.), *The Human Revolution* (Edinburgh University Press, 1989, pp. 212–31). Kristen Hawkes's theory of grandmothering and its effect on human populations is discussed in *Proceedings of the National Academy of Sciences USA* (95: 1336–9, 1998).

Levin and Potapov's *The Peoples of Siberia* (University of Chicago Press, 1964) is an amazing overview of Siberian anthropology – now sadly out of print. Thomas Jefferson's only published book, *Notes on the State of Virginia* (W. W. Norton, New York, 1972), is primarily a collection of facts and figures about the state – although the sections on anthropology are worth reading. Richard Klein reviews much of the material on American archaeology in *The Human Career* (see above). James Chatters's *Ancient Encounters: Kennewick Man and the First Americans* (Simon and Schuster, New York, 2001) describes this exciting archaeological find.

The work by Wallace and Torroni on Native American mitochondrial DNA and multiple waves of migration was reviewed by them in *Human Biology* (64: 271–79, 1992), and by Emoke Szathmary in *American Journal of Human Genetics* (53: 793–9, 1993). Underhill et al.'s paper on the Y-chromosome marker M3 appeared in

Proceedings of the National Academy of Sciences USA (93: 196–200, 1996). Santos et al. and Karafet et al. published their papers on 92R7 and Native American origins in *American Journal of Human Genetics* (64: 619–28 and 64: 817–31, respectively). Joseph Greenberg's work on Native American languages is reviewed in Merritt Ruhlen's *A Guide to the World's Languages*, Volume 1, *Classification* (Stanford University Press, 1987).

8 The Importance of Culture

The epigraph for this chapter is modified from a creation story in Arthur Cotterell's *Encyclopedia of World Mythology* (Paragon, Bath, 1999).

A summarized version of Cook's *Resolution* journal can be found in *The Journals of Captain Cook* (Penguin, London, 1999).

Dame Kathleen Kenyon's book *Digging up Jericho* (Ernest Benn, London, 1957) is her account of the discovery of the origins of the Near Eastern Neolithic. Brian Fagan's account of Neolithic origins is given in his *People of the Earth*, cited above. Cavalli-Sforza and his colleagues' work on the Wave of Advance is summarized in Ammerman and Cavalli-Sforza, *Neolithic Transition and the Genetics of Populations in Europe* (Princeton University Press, 1984), and in Cavalli-Sforza's books cited above. Martin Richards et al.'s work on the mtDNA evidence for a Neolithic expansion was published in *American Journal of Human Genetics* (59: 185–203, 1995), and Semino et al.'s work on the Y-chromosome evidence was presented in their paper in *Science* cited above. David Goldstein and colleagues' work on the spread of Y-chromosome lineages in south-east Asia can be found in *American Journal of Human Genetics* (68: 432–43, 2001). The discussion of negative aspects of the Neolithic transition is taken from several sources, including Fagan (see above), William McNeill's *Plagues and Peoples* (Doubleday, New York, 1976) and *The Cambridge Encyclopedia of Human Evolution* (Cambridge University Press, 1992).

Merritt Ruhlen's *A Guide to the World's Languages* (cited above) and *The Origin of Language* (John Wiley, New York, 1994) and

Charles Barber's *The English Language* (Cambridge University Press, 1993), give general accounts of linguistic classification and the history of linguistics. The *Cambridge Encyclopedia of Language* edited by David Crystal (Cambridge University Press, 1997) is an excellent reference. Cavalli-Sforza's work on genetics and language is reviewed in *The History and Geography of Human Genes* cited above, and the relationship between cultural and biological evolution is examined in greater detail in Cavalli-Sforza and Feldman's *Cultural Transmission and Evolution: A Quantitative Approach* (Princeton University Press, 1981). The search for the Indo-European Homeland is reviewed in Colin Renfrew's *Archaeology and Language* (Cambridge University Press, 1987) and in Jim Mallory's *In Search of the Indo-Europeans* (Thames and Hudson, London, 1989) – both wonderfully engaging books.

Mark Seielstad and colleagues' work on patrilocality and Y-chromosome variation was published in *Nature Genetics* (20: 278–80, 1998), as was the work of Stoneking and colleagues on the matrilocal tribes of northern Thailand (*Nature Genetics* 29: 20–1, 2001).

9 The Final Big Bang

Nationalism and the rise of monolingualism is briefly summarized in Timothy Baycroft's *Nationalism in Europe, 1789–1945* (Cambridge University Press, 1998). The extinction of the world's languages is discussed in David Nettle and Suzanne Romaine's book *Vanishing Voices* (Oxford University Press, 2000). The US census data is available on a US government website (http://www.census.gov/). The statistics on race identity were taken from an article on the census by Steven Holmes in the *New York Times* (3 June 2001).

Index of Pictures

San, Nyae Nyae Pam,
Namibia

San, Nyae Nyae Pam,
Namibia

San, Nyae Nyae Pam,
Namibia

San, Nyae Nyae Pam,
Namibia

San, Nyae Nyae Pam,
Namibia

San, Nyae Nyae Pam,
Namibia

San, Nyae Nyae Pam,
Namibia

San, Nyae Nyae Pam,
Namibia

cave paintings,
Laura, Australia

Aborigine, Cairns,
Australia

Aborigine, Cairns,
Australia

Aborigines, Cairns,
Australia

Aborigine, Cairns,
Australia

Aborigine, Cairns,
Australia

Aborigine, Laura,
Australia

Aborigine, Laura,
Australia

Kazaks, near Semey,
Kazakstan

Azeri, Baku,
Azerbaijan

Dungans, Osh
region, Kyrgyzstan

Kazak, Lake
Markakol, Kazakstan

Kyrgyz, Song-Köl,
Kyrgyzstan

Kyrgyz, Pik Lenina,
Kyrgyzstan

Chukchi, Amguema
Tundra, Russia

Chukchi, Amguema
Tundra, Russia

Chukchi, Amguema
Tundra, Russia

Chukchi, Amguema
Tundra, Russia

Chukchi, Amguema
Tundra, Russia

Chukchi, Amguema
Tundra, Russia

Chukchi, Amguema
Tundra, Russia

Chukchi, Amguema
Tundra, Russia

Chukchi, Amguema
Tundra, Russia

Chukchi, Amguema
Tundra, Russia

Chukchi, Amguema
Tundra, Russia

Navajo, Canyon de
Chelly, Arizona, USA

Picuris, Pueblo,
New Mexico

San Miguel de
Allende, north
central Mexico

Tatras region,
Poland

near Moriarty,
New Mexico, USA

Stonetown, Zanzibar,
Tanzania

Maasai Mara, Kenya

Tokyo, Japan

Russian, near
Almaty, Kazakstan

Puerto Viejo,
Costa Rica

Arab, Bukhara,
Uzbekistan

Page numbers in bold refer to figures